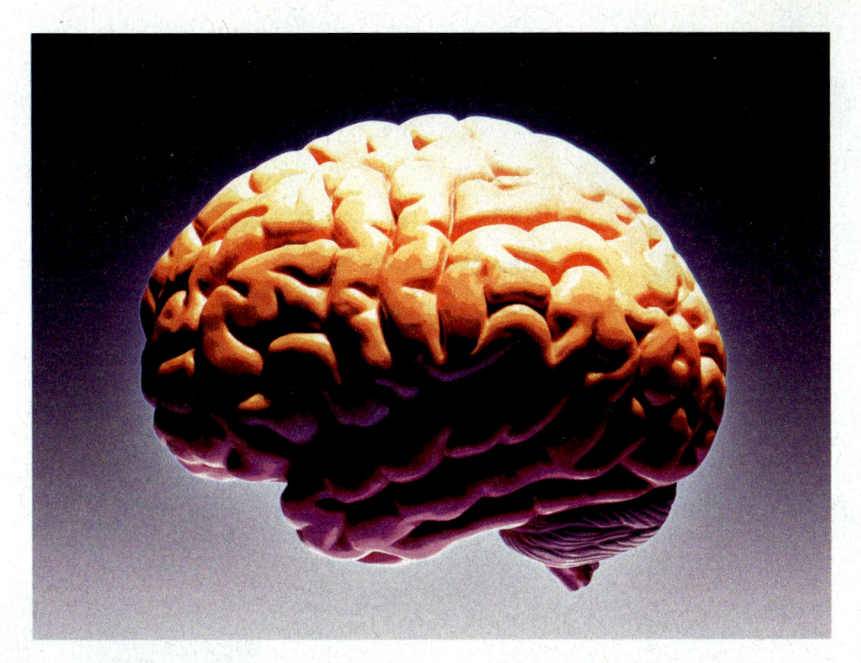

Secrets of Your Brain

U.S.News & WORLD REPORT
usnews.com

Copyright © 2011 by U.S.News & World Report L.P., 1050 Thomas Jefferson Street NW, Washington, DC 20007-3837, ISBN 978-1-931469-44-9. All rights reserved. Published by U.S.News & World Report L.P., Washington, D.C. Printed in the U.S.A. on Future Connection 76 Satin, which contains at least 85 percent recycled fiber, including at least 25 percent post-consumer fiber.

Secrets of Your Brain

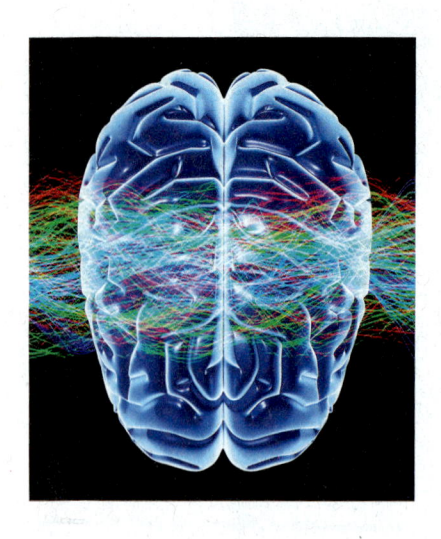

5 Introduction: A Fantastic Journey of Discovery
Scientists have been striving to understand the brain for 2,500 years

CHAPTER ONE
UNDERSTANDING YOUR BRAIN

12 It's One Amazing Machine
A tour of your command center, 100 billion neurons strong

16 Peering Into a Mind at Work
Thanks to new brain imaging technologies, exciting insights are rapidly piling up

20 Inside Your Teen's Brain
That nerve-racking risk-taking? It's a necessary developmental step

23 Rewiring Your Decision-Making
Cognitive shortcuts make choices easy—and sometimes deadly

26 How Optical Illusions Deceive
Artists, magicians, and the military benefit from the brain's limitations in interpreting what people see

32 Straight-A Students, Take Note
Daniel Goleman on why emotional intelligence is key to success

CHAPTER TWO
MYSTERIES OF THE MIND

36 What Dreams Are Made Of
Research is revealing what happens during sleep (and why it matters)

42 Divining the Secrets of the Soul
Scientists and people of faith seek to understand out-of-body experiences

45 The Puzzle of the Psychopath
Studies suggest that the criminal brain is distinctly different

48 The Brain Chemistry of Love
Your heart is racing and you're obsessed? Blame the storm in your head

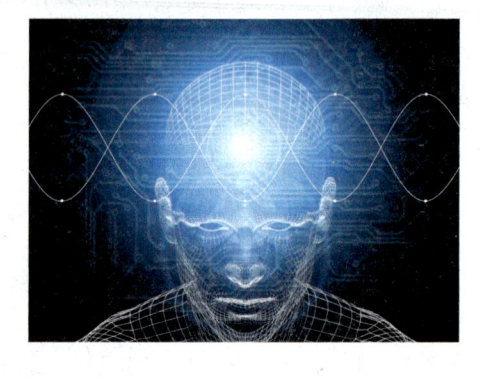

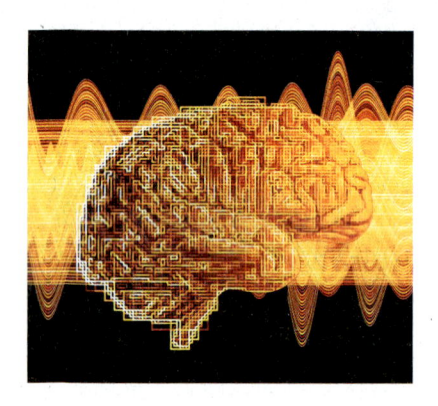

CHAPTER THREE
USING YOUR BRAINPOWER

52 Keeping Your Mind Sharp
One of the best things you can do is get some exercise

57 Making Memories That Stick
These strategies should help your brain do a better storage job

60 The Hyperactivity Advantage
Consider an alternative perspective: that having ADHD is a gift

63 A Different Kind of Learner
How the brains of dsylexic readers process the written word

66 The Epicenter of Pain
The degree of hurt you feel is controlled in your head

68 How to Unleash Your Genius
The creative process involves both art and a good bit of science

CHAPTER FOUR
HEALING THE BRAIN

72 Exploring Medicine's Frontiers
The latest discoveries, from Alzheimer's disease and autism to cancer, concussion, and stroke

79 What if the Gloom Won't Lift?
When drugs don't offer relief, some severely depressed people find that electroconvulsive therapy does

82 Three Patients, Three Operations
Exclusive: Inside the OR at UCSF Medical Center

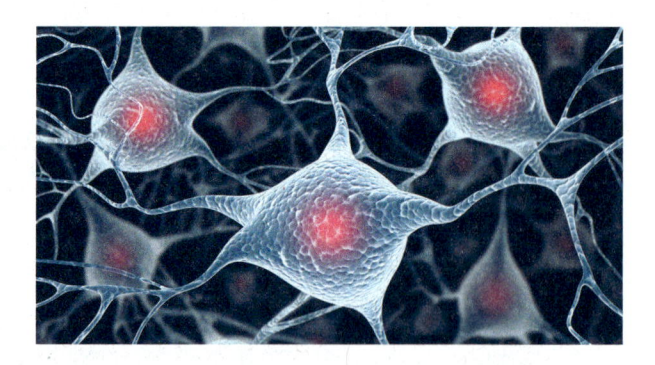

U.S. GOV'T GOLD AT-COST

TODAY - The United States Rare Coin & Bullion Reserve has scheduled the final release of U.S. Gov't Issued $5 Gold Coins previously held at the U.S. Mint at West Point. These Gov't Issued Gold Coins are being released on a first-come, first-serve basis, for the incredible markup-free price of only $155.37 each. This "at-cost" Gov't Gold offer will be available for only a limited time, so do not delay. Call a Sr. Gold Specialist today.

OWN GOV'T ISSUED GOLD COINS

DUE TO STRICT LIMITED AVAILABILITY, TELEPHONE ORDERS WILL BE ACCEPTED ON A FIRST-COME, FIRST-SERVE BASIS ACCORDING TO THE TIME AND DATE OF THE ORDER.

Markup-Free Price of ONLY
$155.37 EACH

If you've been waiting to move your hard-earned money into precious metals, the time is now to consider transferring your U.S. dollars into United States Government Gold. The Gold market is on the move, up more than 400% over the past 10 years - outpacing the DOW, NASDAQ and S&P 500. Call immediately to order your United States Gold Coins direct from our Main Vault Facility, "at-cost", for the amazing price of only $155.37 per coin. Special arrangements can be made for Gold purchases over $50,000. Order your Gold today!

1 – Gov't Issued Gold Coin	$	155.37
(PLUS INSURANCE, SHIPPING & HANDLING $31.00)		
5 – Gov't Issued Gold Coins	$	776.85
(PLUS INSURANCE, SHIPPING & HANDLING $31.00)		
10 – Gov't Issued Gold Coins	$	1,553.70
(PLUS INSURANCE, SHIPPING & HANDLING $36.00)		

DUE TO MARKET FLUCTUATIONS, AT-COST PRICES ARE VALID FOR A MAXIMUM OF 30 DAYS FROM AD PUBLICATION DATE. SPECIAL AT-COST OFFER IS STRICTLY LIMITED TO ONLY ONE LIFETIME PURCHASE OF 10 AT-COST COINS (REGARDLESS OF PRICE PAID) PER HOUSEHOLD, PLUS SHIPPING AND INSURANCE.

Distributor of Government Gold. Not affiliated with the U.S. Government.

CALL TOLL FREE (24 Hours A Day, 7 Days A Week)
1-888-514-5008
MASTERCARD • VISA • AMEX • DISCOVER • CHECK

UNITED STATES
RARE COIN & BULLION RESERVE
Distributor of Government Gold. Not affiliated with the U.S. Government.

Vault No. US42-15537

Coins enlarged to show detail.

No one, including the United States Rare Coin and Bullion Reserve, can guarantee a Gold Coin's future value will go up or down. © 2011 United States Rare Coin and Bullion Reserve

Advertisement

Tips on Leading a Long & Healthy Life

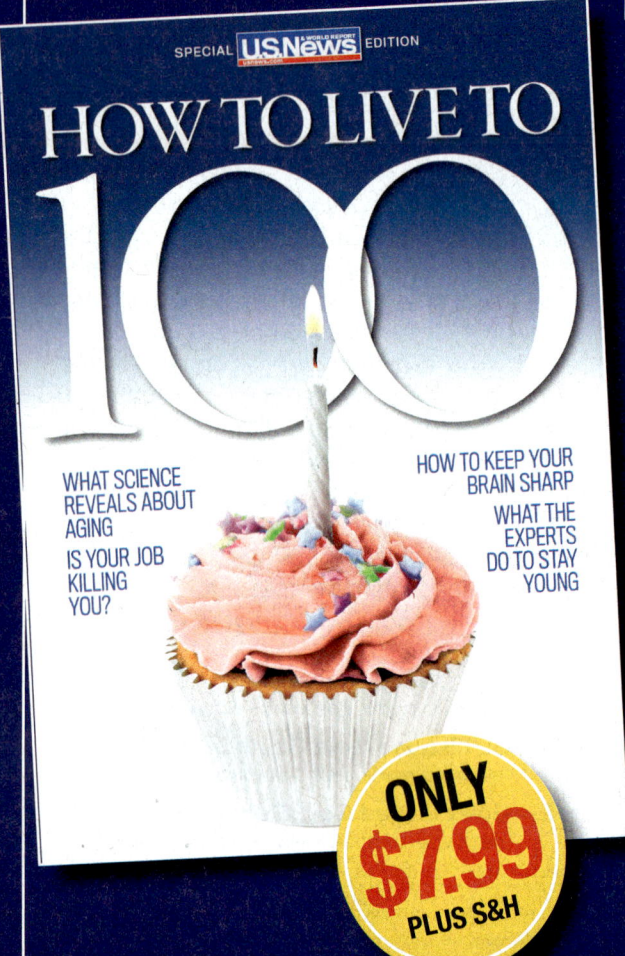

SPECIAL U.S.News & WORLD REPORT EDITION

HOW TO LIVE TO 100

WHAT SCIENCE REVEALS ABOUT AGING
IS YOUR JOB KILLING YOU?
HOW TO KEEP YOUR BRAIN SHARP
WHAT THE EXPERTS DO TO STAY YOUNG

ONLY $7.99 PLUS S&H

How well you age has plenty to do with your genes and luck. But as the latest science on longevity shows, you have a lot more control than you might think. In this special edition, U.S. News & World Report explores the many actions you can take now that should help you hit the century mark still independent, vital, and strong.

Find out how your diet and regular exercise can protect you, how that spare tire around your middle endangers your longevity (and how to shed it), how to sustain a healthy love life, and how to keep your brain fit.

PLUS:
- A 10-week workout plan designed for people over 40
- One Minnesota town's health makeover
- Special section: Vacation getaways voted tops by older travelers

ORDER NOW AT
www.usnews.com/Liveto100 or call 1-800-836-6397

U.S.News & WORLD REPORT
usnews.com

SECRETS OF YOUR BRAIN

A Fantastic Journey of Discovery

On the table was a corpse, surrounded by knives and saws, turpentine and syringes. An audience of astronomers, philosophers, scientists, and priests had gathered at Beam Hall in Oxford, England, in 1662, to witness a rare dissection. In the past, anatomists had always separated the brain in slices, taking so long to study each layer that lower portions of the organ decayed, making it impossible to form an accurate idea of its internal structures. On this day, the eminent Oxford physician Thomas Willis took a new approach. He opened the skull and lifted the entire brain from its cavity at once. The doctor pointed out the cerebrum, with its two separate hemispheres. In the process, he made a discovery: a circle of arteries supplying blood to the brain. It would subsequently be called the "Circle of Willis" in his honor. Though the autopsy would prove to be a turning point in neuroscience, Willis initially met with resistance. He was challenging entrenched beliefs that would only harden over the next 100 years. Willis argued that different parts of the brain performed separate jobs, each with its own line of blood supply. Critics were unimpressed. After all, said Danish anatomist Nicolaus Steno, "How can he be so sure?"

The history of neuroscience has been one of many twists and turns. It's a story of overcoming intellectual prejudice, scant resources, and the eccentricity of a wild cast of characters. Yet today, researchers could not study or treat ailments as varied as Parkinson's, concussions, tumors, mental illness, and Alzheimer's without the extraordinary persistence, imagination, and brilliance of scientists like Willis. Progress was never linear, and it makes for a fascinating tale.

During the Golden Age of Greece, physicians debated whether the heart or the brain served as the body's control center, and whether demons or anatomy dictated outcomes. Hippocrates, whose instruction "First, do no harm" still guides medical ethics, believed that illness came from an imbalance

Andreas Vesalius's 1543 drawing of the brain with the two hemispheres exposed.

SPECIAL EDITION 5

This illustration shows the lines of phrenology, a debunked theory that personality traits could be determined by reading a skull's shape, bumps, and depressions.

in nature. Born on the Aegean island of Cos around 460 B.C., he was the first doctor to suggest that epilepsy was not a punishment for earthly sins but resulted from a variance in the brain. Instead of the popular "devil" theory proposed by other healers, he suggested that excess blood might be building up in the skulls of some brain injury patients. For these sufferers he would bore a hole in the head to allow drainage and to re-establish the body's balance of its four humors, whose corresponding elements are air, fire, earth, and water. The father of Western medicine may not have understood why easing cranial pressure would work, says Stanley Finger, editor of the *Journal of the History of the Neurosciences*, but by doing this procedure he saved lives.

But few physicians dared challenge orthodox views of their time. Not until the first millennium did facts begin to gain on myths. Born in A.D. 129, Galen of Pergamum (now Bergama, Turkey) treated the head wounds of Roman gladiators and was considered the greatest physician of antiquity. The Greek philosopher Aristotle had theorized that the brain served merely as a cooling station for blood circulating in the body, but more than 400 years after his death, Galen wondered. He gave a series of dissections that would guide science for centuries. In a demonstration before a crowd of physicians, philosophers, and politicians, he opened the skull of a just-killed animal and found the brain still warm, refuting Aristotle.

In another experiment that Princeton University psychologist Charles Gross has described as "famous in its own time and for centuries later," Galen dissected a squealing pig that was strapped to a table. When Galen cut the nerves linking the larynx to the brain, the struggling continued, but the squealing stopped. Here was proof, said the Roman Empire's top medical expert, that the nerves connect to the voice not from the heart but from the brain.

Human autopsies, however, were frowned upon by the Roman citizenry at the time. In the face of this disapproval, Galen had to rely on animal dissections, from which he drew conclusions on human anatomy. This led to his making many mistakes. For example, he assumed that a network of blood vessels found at the base of the brain of hoofed animals also existed in humans. These misconceptions would become settled science for centuries to come. "Like a bear approaching the cold of winter, the spirit of scientific inquiry slipped into hibernation," Finger wrote in *Minds Behind the Brain*, his biographical account of the field's pioneers.

Not until the creative explosion of the Renaissance would science revive. A brash, 29-year-old Flemish anatomist, Andreas Vesalius, defied church prohibitions by covertly dissecting human cadavers, some stolen from the gallows. In 1543, he published a seven-volume work, *On the Fabric of the Human Body*, that would later be called "the most accurate book on human anatomy up to that time." Correcting many errors in Galen's work, Vesalius became the first scientist to publish illustrations of regions of the brain like the thalamus and the cerebellum.

Leonardo's secrets. Ironically, the honor of upending Galen might have fallen to Leonardo da Vinci decades earlier, but for the artist's obsession with secrecy. One of the greatest minds of a great age, da Vinci is remembered for his famous painting, *Mona Lisa*, and for dozens of futuristic inventions that he mapped out in his notebooks. Born in 1452, the Italian was permitted, as a successful artist, to dissect corpses at local hospitals, the better to draw human bodies. And draw he did—anatomical depictions of astonishing accuracy. Driven to understand the anatomy that informed his paintings, da Vinci poured melted wax into the ears of a dead ox. After the wax hardened, he peeled away the surrounding tissue, delineating the contours of the brain. But da Vinci kept his drawings private. Following his death in 1519, they remained unpublished for centuries, depriving the world of his remarkable illustrations and anatomical insights.

A century after Vesalius began rescuing neuroscience from its long hibernation, Thomas Willis made his important discoveries on the brain's anatomy while treating patients during a 1657 meningitis

DID YOU KNOW...

When Albert Einstein died on April 18, 1955, his brain was removed and preserved by Thomas Harvey, the pathologist who performed the autopsy. Harvey kept the brain for more than 40 years, occasionally carving off pieces and giving them to prominent neuroscientists for research.

SOURCES: *DRIVING MR. ALBERT* BY MICHAEL PATERNITI; *THE OTHER BRAIN* BY DOUGLAS FIELDS

SECRETS OF YOUR BRAIN

epidemic. Curious about why his patients were losing their rational minds, he persuaded families of those who died to let him investigate the cause, thus circumventing continuing church edicts against human dissections. What he discovered was that all the brains of those felled by the illness were coated with a thick bloody mass, suggesting that intellect was housed in the brain. Intrigued by his findings, Willis would devote the rest of his life to studying the nerves that connected the brain to the body, a field he named "neurologie."

By the 1800s, fascination with medical experimentation had moved from the morgue to the public square. In 1818, novelist Mary Shelley published *Frankenstein*, her classic tale of a doctor who brings a corpse to life by infusing bioelectricity into its brain. Critics wondered if the book was a metaphor for scientists gone mad in the lab. And, in fact, serious research was often undermined by pseudoscience. Beginning in 1796, the German neuroanatomist and physiologist Franz Joseph Gall traveled widely through Europe, spreading his ideas. Gall theorized that the brain was divided into 27 different "organs" controlling specific areas of localized function, like speech. He also believed that a specialist could read the bumps, depressions, and shape of a skull to determine personality traits and abilities. Bump reading or phrenology, which was later discredited, became all the rage across Europe and then spread to America.

Many of Gall's peers scoffed at his theories, and French physiologist Jean Pierre Flourens set out, in 1825, to disprove them. In a series of experiments, he removed parts of the brains of rabbits and pigeons to assess how these surgical losses affected them. He noticed that only when he removed large portions of the cerebral hemispheres were behavior and movement impaired. If he removed the brainstem, the animals died. Still, because he saw that removing smaller sections of the cortex did not drastically alter behavior, Flourens concluded that the brain worked as a single unit. On this point, history would prove Gall's general idea about localized function as being closer to reality. What the German misunderstood was the brain's internal structure.

On Sept. 13, 1848, 25-year-old Phineas Gage, a construction foreman for the Rutland & Burlington Railroad, had an accident that would further animate debate about the brain's function. While preparing a roadbed near Cavendish, Vt., Gage used a large tamping rod—over 3 feet long and an inch in diameter—to compact blasting powder and sand in a hole drilled into rock. An explosion sent the rod roaring out of the rock, and it rammed through the side of Gage's face, entering at his left cheek, passing through the back of his left eye and frontal lobe, and exiting from the top of his head, spewing blood and brain on the road. After the accident, Gage recovered all his physical faculties. But his personality changed completely. Once a responsible gentleman, he now turned coarse and vulgar. His physician, John Martyn Harlow, concluded that the psychological changes in Gage's personality were a result

Andreas Vesalius defied church prohibitions by COVERTLY DISSECTING human cadavers

of damage to his brain. Though Harlow published his findings, the scientific community met them with skepticism.

In 1861, another doctor would further galvanize debate about how the brain functions. At the Bicêtre Hospital for men outside Paris, two patients consulted neurosurgeon Pierre Paul Broca. Their case files gave weight to Harlow's theory that brain injury could affect behavior and suggested the importance of the left frontal lobe in controlling language. A 51-year-old patient named Leborgne, epileptic since childhood, came to Broca able to speak only one word, "tan." Within days the patient was dead. In an autopsy, Broca discovered a lesion on the man's left frontal lobe. Soon after, an 84-year-old stroke victim named Lelong visited the doctor. Lelong also had been reduced to speaking just a few words, including *oui* and *non*. After his death, the autopsy revealed damage to his lateral front lobe, in an area close to that in Leborgne's brain. An impressed Société d'Anthropologie hailed these findings, and the section of the brain responsible for fluid speech came to be known as Broca's area.

Mind-body connections. This breakthrough, tying a region of the brain to a specific role for the first time, encouraged efforts to probe these mind-body connections. In October 1889, Santiago Ramón y Cajal, a Spanish scientist and artist, attended a meeting of the German Anatomical Society at the University of Berlin to share an exciting discovery. Having

Galen (above), antiquity's greatest physician; Hippocrates (left, top), who treated brain injuries by boring holes in patients' skulls; and Leonardo da Vinci, who hid his anatomical drawings.

improved the staining techniques used to examine tissue under a microscope, he reported that nerves were not, as earlier believed, welded together in networks or baskets leading from the cerebral cortex, the spinal cord, or the olfactory bulb. Instead, Cajal announced, the nerves were independent actors. Later scientists came to understand that these neurons signaled one another across the gaps or "synapses" that divided them. (Researchers now believe some diseases of the nervous system may result from disorders of neuronal signaling molecules.)

Today, neurologists believe rehabilitation is as key to **TREATING BRAIN TRAUMA** as surgery

Cajal's discoveries marked a more sophisticated chapter in brain research. A year after the Berlin meeting, Russian Ivan Petrovich Pavlov was hired to organize and direct the physiology department at the St. Petersburg Institute of Experimental Medicine. Over the next four decades, Pavlov conducted animal experiments that linked his name to their behavior, winning the 1904 Nobel Prize for his work on the flow of gastric juices. In his most famous procedure, Pavlov conditioned dogs to salivate at the sound of a bell after they were trained to associate its ring with food. The Society for Neuroscience now cites this research in discussing post-traumatic stress disorder, explaining that "some traumatic events like natural disasters, war, violent attacks, and serious accidents can cause lasting associations and effects that interfere with daily life."

Perhaps no group of patients suffered more from the medical profession's ignorance than the mentally ill. Locked away in asylums beginning in the 1700s, they were chained to walls and kept in dungeons. Stories of patient mistreatment are rampant and involve even enlightened men of science in the 20th century.

Henry Cotton was a respected psychiatrist who took some steps to improve patient care when he became medical director of the New Jersey State Hospital at Trenton in 1907. A progressive thinker, he instituted daily staff meetings on each patient and tried to end the practice of keeping inmates chained to their beds. But he also believed, like Hippocrates, that mental illness was caused by infection. To test his theory, he surgically removed the teeth, sinuses, colons, and sexual organs of patients suffering from hallucinations.

Over succeeding decades, doctors sought better ways to treat the severely mentally ill. One approach would prove highly controversial. It involved cutting connections to and from the frontal lobes of the patient's brain. In 1949, the Nobel Prize for medicine was awarded to Portuguese physician António Egas Moniz for discovering the therapeutic value of the surgery. Called a lobotomy, it would become a standard procedure for more than two decades. Over time, though, outrage would grow over the operation, which blunted the personality and made patients apathetic and childlike.

Searching for new treatment options, researchers turned to pharmaceuticals. Australian psychiatrist John Cade reported in 1949 that lithium could treat bipolar disease. By 1954, the first antipsychotic medicines would reach the market, followed by antidepressants in the late 1950s and benzodiazepines (for anxiety) in 1961. All offered hope for patients seeking to live normal lives.

Improving technology. Virtually all the breakthroughs of the modern era would not have been possible, though, without major technological advances. In 1895, a physicist named Wilhelm Röntgen was studying the effects of electrical currents on vacuum tubes at a university lab in Würzburg, Germany. In a darkened room, Röntgen placed a cathode-ray tube in a black cardboard box. When he sent the current through the tube, he noticed a nearby screen glow with fluorescent light. He soon showed his discovery to his wife, Anna Bertha, and imaged her hand. When she saw her bones illuminated, she reportedly exclaimed, "I have seen my death!" Röntgen called the pictures X-rays, the X a symbol of his uncertainty, and in 1901 he was awarded the first Nobel Prize for physics. His device revolutionized medicine, opening the door to a cascade of new diagnostic tools such as EEGs, PET scans, and MRIs (story, Page 16).

Future advances in genetic research and molecular biology offer further promise. "The ability to sequence the whole genome of individual patients—to understand brain diseases from autism to schizophrenia, to untangle the genetic from the physical—it's all very exciting," says Jonathan Pevsner, who runs a bioinformatics lab at the Kennedy Krieger Institute in Baltimore and studies the molecular basis of childhood brain disorders.

Perhaps the most unexpected development in the history of neuroscience was the discovery that

DID YOU KNOW...

Everything people see is processed from the eyes through the brain's visual cortex. One device currently under development is the retinal implant, which uses electrodes to transfer information directly into the brain, allowing someone to "see" while bypassing the normal path from the eyes.

SECRETS OF YOUR BRAIN

New treatments aided Rep. Gabrielle Giffords's recovery after her traumatic brain injury.

the brain has some ability to overcome trauma. The concept, called neuroplasticity, owes its discovery to a Mexican immigrant to the United States whose sons both became doctors. In 1958, 65-year-old Pedro Bach-y-Rita suffered a stroke, leaving one side of his body paralyzed and compromising his ability to speak. Treated by his son George, Pedro managed to recover, even going mountain climbing. After his death, his son Paul, a neuroscientist, learned from the autopsy that the damage to one side of Pedro's brainstem remained. Somehow, the brain had reorganized itself. Today, neurologists believe rehabilitation is as key to healing brain trauma as surgery. Countless people now benefit not just from new life-saving treatments, but also from better therapies that produce patient outcomes all but unthinkable a decade ago.

New protocols. These developments aided Arizona Rep. Gabrielle Giffords's remarkable recovery from traumatic brain injury in the days after Jan. 8, 2011, when she was shot in Tucson. To Geoffrey Manley, professor of neurosurgery at the University of California–San Francisco, her recovery, while "certainly not a miracle," rather "reflects progress in traumatic brain injury care." Ten years ago, he says, clinician-scientists focused on ways to improve the outcomes for patients with these devastating injuries. A number of these updated protocols were followed by the University Medical Center in Tucson, where Giffords was first taken.

For example, she received specialized care within 40 minutes of the shooting. (An impressive 88 percent of patients with severe injuries in Arizona are within an hour of a top-level trauma center, compared to 61 percent in neighboring New Mexico.) To relieve Giffords's brain swelling, surgeons borrowed the ancient practice of removing part of the skull. The procedure, called decompressive craniectomy, was refined by U.S. military surgeons in Afghanistan and Iraq and is now routinely used in leading trauma centers. Giffords, who was placed in a medically induced coma, also benefited from continued, careful monitoring of the pressure within her skull, preventing further injury. Finally, as soon as she was sufficiently recovered, she was transported to a top rehabilitation center in Houston.

Though improved care has led to better outcomes, Manley laments that there are no national standards for trauma centers and inadequate funding to study drug treatments that could control swelling without removing the skull. "Giffords was lucky," he says. "Tucson was really well organized."

That she survived the trauma with a chance to rebuild her life owes something to discoveries of the past, no matter how halting their course. That doctors can only guess how full such trauma patients' final recovery will be speaks to how much is yet to be learned. After 2,500 years of speculation, the brain remains what it always has been—an awe-inspiring mystery. ●

By Johanna Neuman

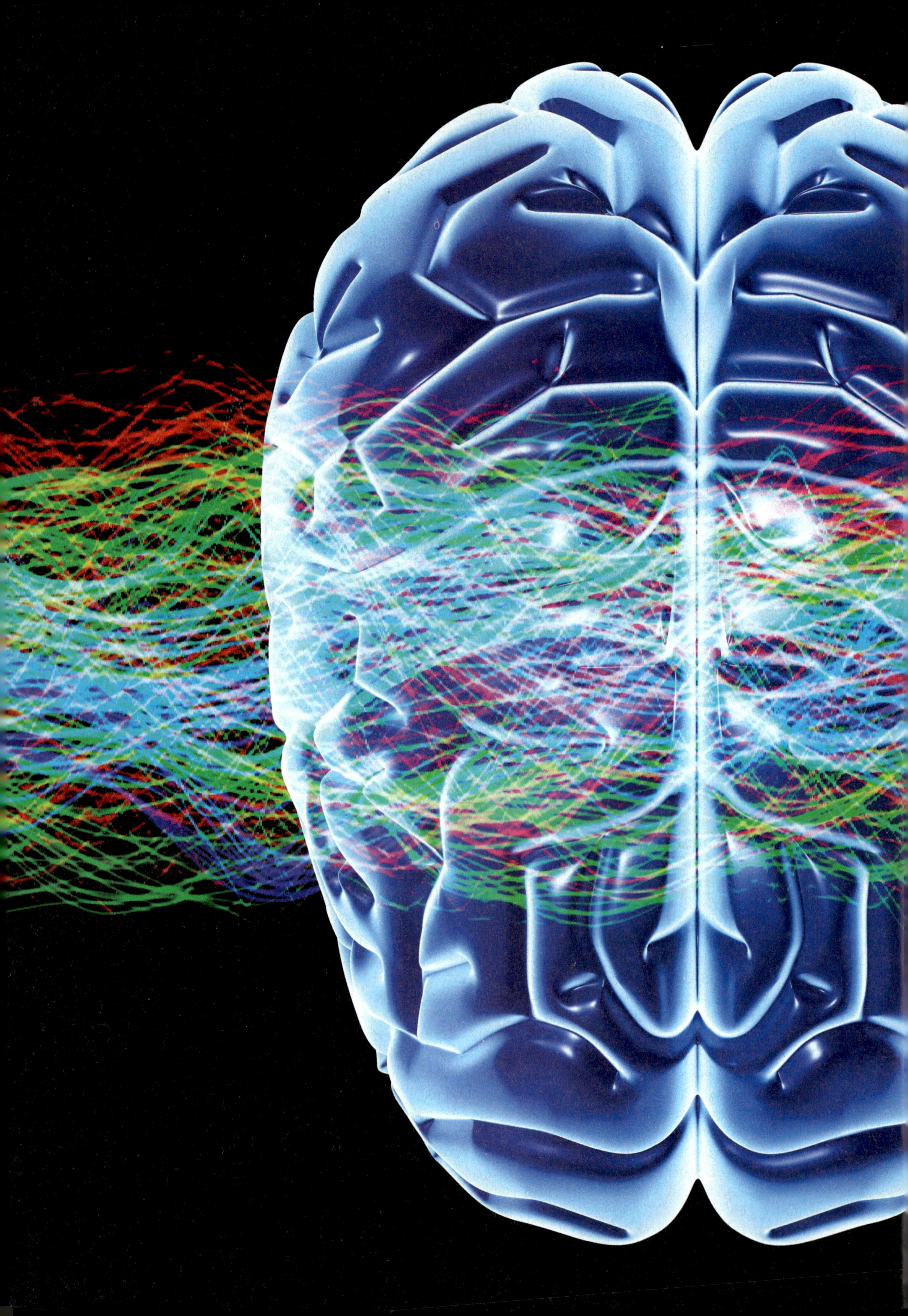

1.
Understanding Your Brain

A primer on all the amazing hardware **12**

What the newest imaging technology reveals **16**

Inside your teenager's baffling head **20**

The cognitive shortcuts that make decisions easy (and sometimes disastrous) **23**

How the art of optical illusion is used to delight and to deceive **26**

Why emotional intelligence matters as much as book smarts **32**

UNDERSTANDING YOUR BRAIN

It's One Amazing Machine

True, it might not look like much to the naked eye: 3 pounds of pink-gray tissue, sealed in liquid, and encased in a tough white membrane resembling the inner skin of a pomegranate. But this unassuming blob, the brains of the operation that makes you uniquely you and people human, represents to many researchers the final and most thrilling frontier in biological science.

Research in the last 50 years has vastly expanded knowledge of the brain's structure and function, but many questions remain. How, exactly, does the brain's hardware store and retrieve memories? What causes devastating neurological disorders such as epilepsy, Alzheimer's, schizophrenia? On a more existential level, "we still can't explain the relationship between the brain and consciousness. The old mind-brain problem," says John Nolte, a neuroanatomist at the University of Arizona. "That's one

Compared to other animals, humans have an extremely LARGE CORTEX– thus their superior intelligence

I'm not sure we're ever going to figure out."

Physically speaking, it's now known that some 100 billion nerve cells, or neurons, are powering a person by the time the brain is fully formed in the early 20s. Just as peripheral nerves in the arms and legs sense pain or tell a muscle to move, the neurons receive and send signals that underlie thoughts, actions, and feelings of love or anger or fear. Extensions from neuron cell bodies convey electrical signals throughout the nervous system; dendrites carry relatively slow signals toward the cell body, and axons move faster impulses away, over longer distances. The neurons connect at specialized junctions, or synapses, where the signals convert to chemical messages that travel between cells. This dense network forms an unparalleled information-processing center that computational scientists dream of reproducing. "People often compare the brain to a computer, but it works much differently," says Nolte. For one thing, neurons don't operate in a strictly digital manner—on or off, yes or no. Rather, they accumulate inputs from other neurons, and send information along only when the information reaches a critical threshold. Moreover, the brain can lose neurons over time through injury or aging, make new connections during childhood and later in life, and even grow new cells, changing the structure of the network.

For most of the past century, since scientists discovered the electrical properties of neurons, exploration has focused on the flashy cells that seemed to do all the work. The glial cells that make up 85 percent of the brain were considered to be mere scaffolding for the neurons. But in just the last 20 years, the glia's truly vital role has become apparent. They synthesize and regulate neurotransmitters such as glutamate and serotonin, the chemicals that neurons use to communicate with each other and that are so key to the way people feel. They also help guide neuron development and synapse formation. "We overlooked glia for a long time because we didn't have the tools to see what they were

Inside the Command Center

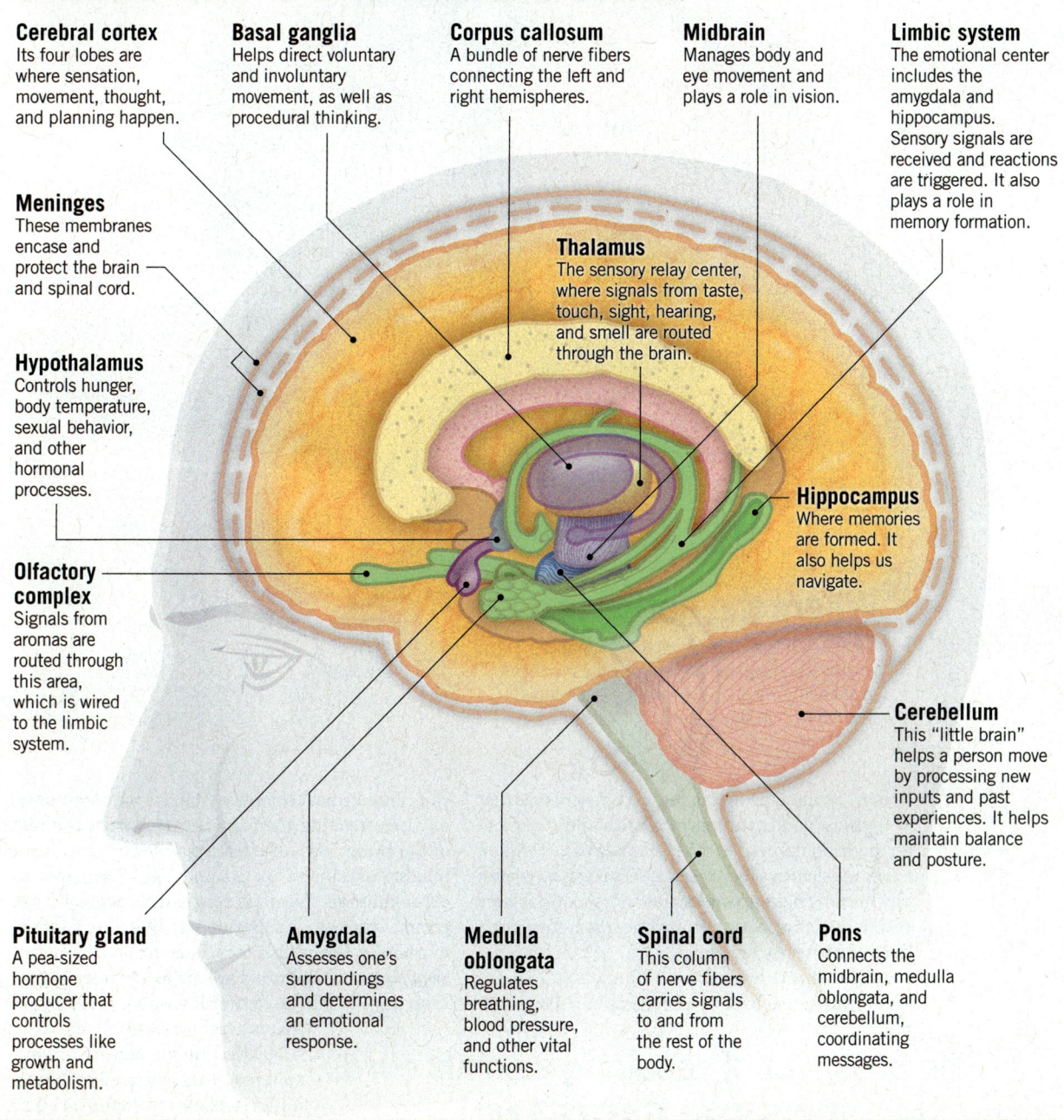

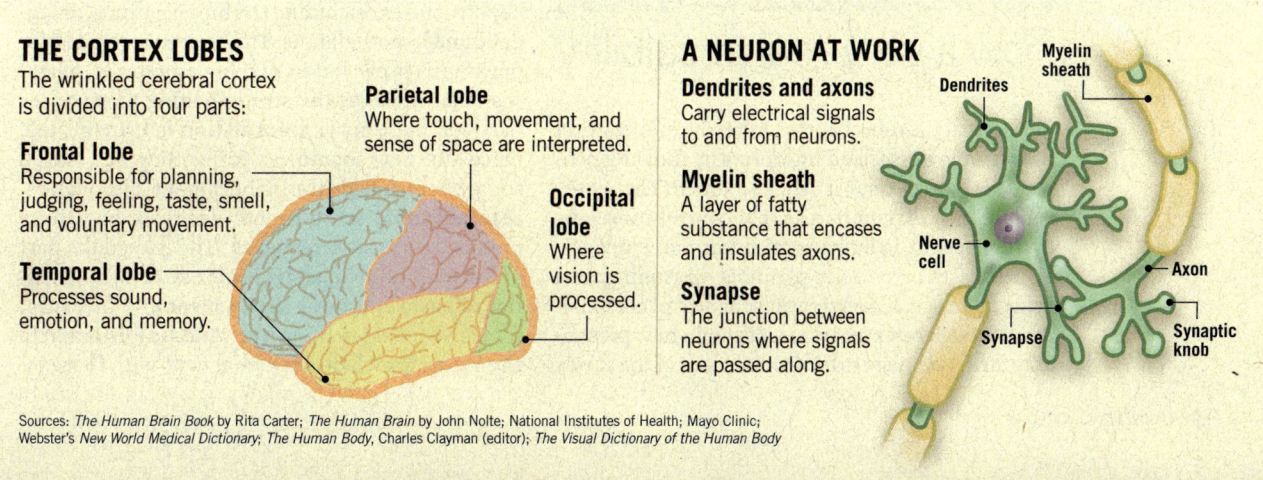

Cerebral cortex
Its four lobes are where sensation, movement, thought, and planning happen.

Basal ganglia
Helps direct voluntary and involuntary movement, as well as procedural thinking.

Corpus callosum
A bundle of nerve fibers connecting the left and right hemispheres.

Midbrain
Manages body and eye movement and plays a role in vision.

Limbic system
The emotional center includes the amygdala and hippocampus. Sensory signals are received and reactions are triggered. It also plays a role in memory formation.

Meninges
These membranes encase and protect the brain and spinal cord.

Thalamus
The sensory relay center, where signals from taste, touch, sight, hearing, and smell are routed through the brain.

Hypothalamus
Controls hunger, body temperature, sexual behavior, and other hormonal processes.

Hippocampus
Where memories are formed. It also helps us navigate.

Olfactory complex
Signals from aromas are routed through this area, which is wired to the limbic system.

Cerebellum
This "little brain" helps a person move by processing new inputs and past experiences. It helps maintain balance and posture.

Pituitary gland
A pea-sized hormone producer that controls processes like growth and metabolism.

Amygdala
Assesses one's surroundings and determines an emotional response.

Medulla oblongata
Regulates breathing, blood pressure, and other vital functions.

Spinal cord
This column of nerve fibers carries signals to and from the rest of the body.

Pons
Connects the midbrain, medulla oblongata, and cerebellum, coordinating messages.

THE CORTEX LOBES
The wrinkled cerebral cortex is divided into four parts:

Frontal lobe
Responsible for planning, judging, feeling, taste, smell, and voluntary movement.

Temporal lobe
Processes sound, emotion, and memory.

Parietal lobe
Where touch, movement, and sense of space are interpreted.

Occipital lobe
Where vision is processed.

A NEURON AT WORK

Dendrites and axons
Carry electrical signals to and from neurons.

Myelin sheath
A layer of fatty substance that encases and insulates axons.

Synapse
The junction between neurons where signals are passed along.

Sources: *The Human Brain Book* by Rita Carter; *The Human Brain* by John Nolte; National Institutes of Health; Mayo Clinic; Webster's *New World Medical Dictionary*; *The Human Body*, Charles Clayman (editor); *The Visual Dictionary of the Human Body*

Where the Action Is

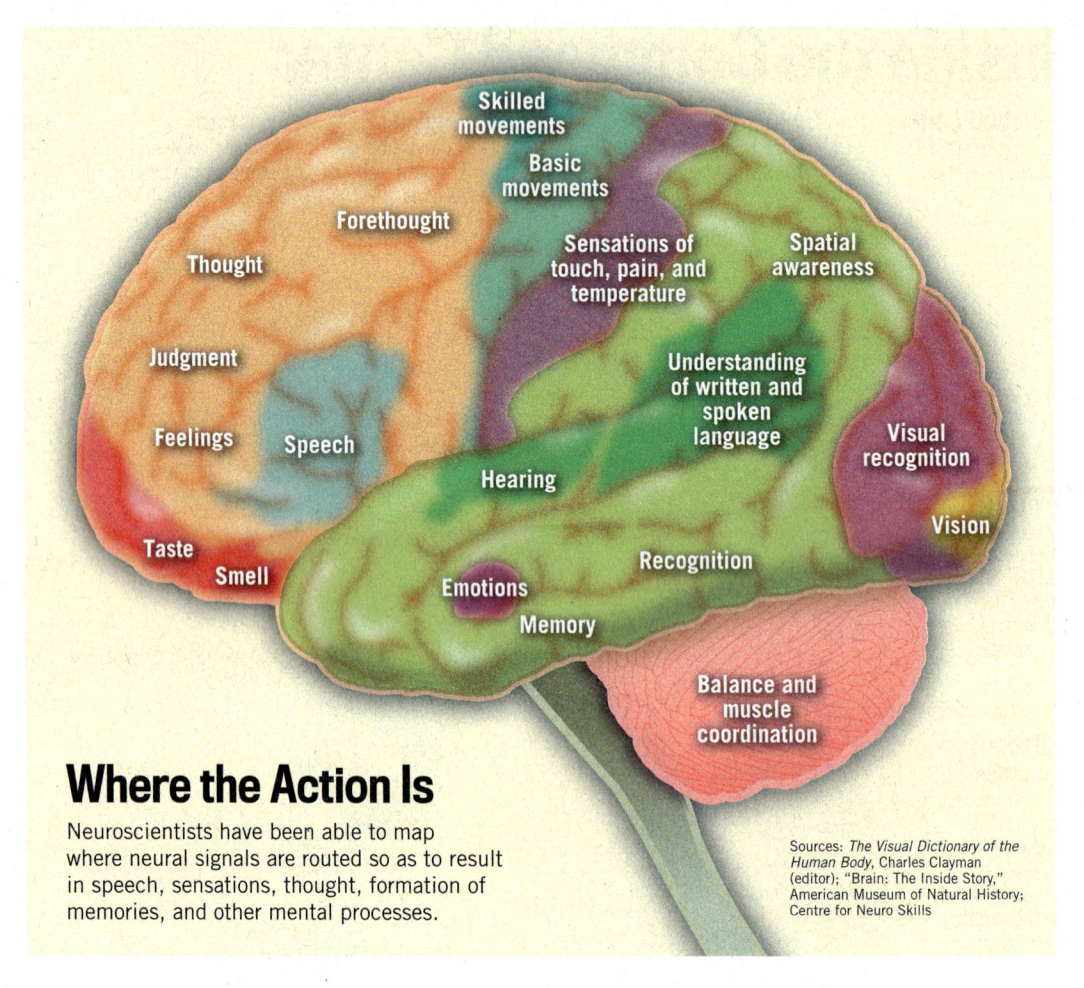

Neuroscientists have been able to map where neural signals are routed so as to result in speech, sensations, thought, formation of memories, and other mental processes.

Sources: *The Visual Dictionary of the Human Body*, Charles Clayman (editor); "Brain: The Inside Story," American Museum of Natural History; Centre for Neuro Skills

doing," explains Douglas Fields, a neuroscientist at the National Institutes of Health and author of a recent book on glia, *The Other Brain*. Using a calcium-sensitive dye, researchers can now watch glia literally light up under the microscope as they respond to stimulation from a nearby neuron or a dose of neurotransmitter.

All vertebrate brains, from humans to fish, include the same basic parts (graphic, Page 13).

The prospect of repairing a spinal cord, once AN IMPOSSIBLE DREAM, is now a very real possibility

The wrinkly cerebral cortex takes in information through the senses and interprets it, allowing people to see, hear, taste, smell, and touch. The cortex is home also to motor areas that control conscious movement and, in humans, it houses the complicated processes involved in planning, reasoning, and using language. It also serves as the seat of personality. Although researchers can't explain how personality forms, damage to the frontal lobes of the cortex can cause radical changes, which is why lobotomy, or disconnecting the frontal cortex from the rest of the brain, was an effective (if brutal) means of making psychiatric patients docile. Compared to other animals, humans have an extremely large cortex—thus their superior intelligence. "In most mammals, brain size correlates to body weight," says Nolte. "Our brains are six to 10 times bigger than they should be for our body size, and most of that is cerebral cortex."

Nestled inside the two hemispheres of the cortex, like a walnut in its shell, is a system of other key structures, including the hippocampus, amygdala, and hypothalamus. The crescent-shaped hippocampus is crucial in making memories, a process that involves the strengthening of synapses between neurons as information is transmitted. Once the new memories form, they are stored somewhere outside the hippocampus, although researchers don't know exactly where, or how those memories are later retrieved. The amygdala, part of a circuit known as the limbic system, is essential to basic emotions such as fear and pleasure; it receives sensory input and sends information to the cortex, which decides what to do with those inputs, and the hypothalamus, which controls primal drives and reflexes such as the fight-or-flight reaction. Strong connections between the amygdala and the hypothalamus help explain why witnessing a car crash, say, can send the heart racing and cause the skin to sweat. The limbic system is also responsible for the rush of joy that follows sex, chocolate, or a cigarette.

Whereas the cortex is responsible for conscious movement, the wrinkled cerebellum at the bottom rear of the brain memorizes repeated patterns and builds the circuits that turn learned dance steps or swimming motions, say, into automatic skills that require no thought. The cerebellum also helps manage equilibrium.

The brainstem, a fibrous mass that connects to the spinal cord, consists mostly of nerve tracts that move information in and out of the brain, and it controls the body's purely automatic essential functions such as breathing and circulation. The midbrain, which is the uppermost region of the brainstem, is involved in movement and includes the dopamine-producing region that gets damaged in Parkinson's disease. The pons, situated just below the midbrain, controls sleep and consciousness. The lowest region of the brainstem, the medulla, connects to the spinal cord and controls breathing, heart rate, blood pressure, and vomiting.

Historically, psychologists and physicians studied which functions occurred where by cutting out small parts of animals' brains and observing what happened, or by examining people with localized brain injuries. The role of the hippocampus in forming memories of new experiences, for example, was discovered by researchers observing Henry Gustav Molaison, an epilepsy patient who suddenly became unable to make new memories after surgeons in 1953 removed sections of his brain that included his hippocampus. The treatment did help control his seizures. But while Molaison could recall his childhood and was able to hold on to new memories briefly, he could no longer turn these immediate records into longer-term records.

Rapidly advancing imaging technology is now providing such insights much more efficiently (story, Page 16). Still, "we've just scratched the surface," says Fields. The hope is that accumulating knowledge will lead to major medical advances. The prospect of repairing a damaged spinal cord, for example, was once an impossible dream; now it's a very real possibility. "We used to think of brains as things that, if they broke, you couldn't fix them," says Nolte. That's simply no longer true. •

By Katherine Leitzell

The Incredible Shrinking Brain

Brain volume starts diminishing in the 20s and accelerates so that it is shrinking by around 0.5–1 percent a year after age 60 on average. Not all brain regions shrink uniformly: The hippocampus tends to shrink faster than the frontal cortex; some areas, like the primary visual cortex, don't shrink at all. Other changes include:

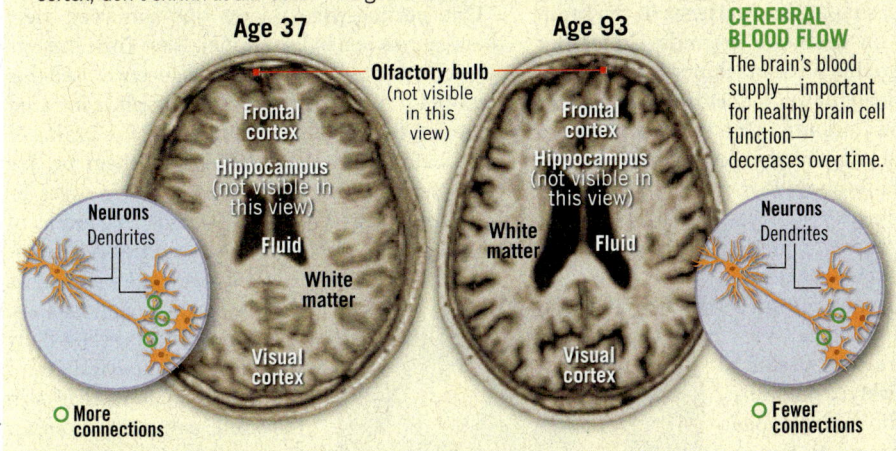

CEREBRAL BLOOD FLOW The brain's blood supply—important for healthy brain cell function—decreases over time.

CAN YOU HEAR ME NOW? Connections between brain cells decline, as dendrites retract in several brain regions including the prefrontal cortex and hippocampus.

A FAILURE TO COMMUNICATE? White matter, which connects various brain regions, becomes less dense in the 50s. As it deteriorates, regions become less coordinated with each other.

GOOD NEWS. The hippocampus and olfactory bulb, which processes smell, continue to produce new neurons. Not all survive to become part of the working brain. But in animal studies, physical exercise and mental stimulation increased brain cell survival rates.

Sources: Brain scans: Randy Buckner—HHMI at Harvard University; inset images: Cabeza R, Anderson ND, Locantore JK, McIntosh AR (2002) Aging gracefully: Compensatory brain activity in high-performing older adults. Neuroimage 17:1394-1402

DRAWING ON EXTRA RESOURCES

Older brains use different areas than younger brains to perform some tasks. One study involving a memory task showed:

Younger people used only one side of the brain.

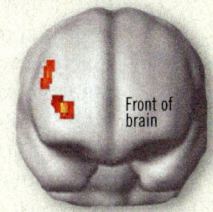

Older people often used both sides. This shift may show the brain adapting to slower processing speed and less efficient neural networks.

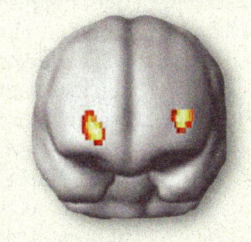

UNDERSTANDING YOUR BRAIN

Peering Inside a Mind at Work

THANKS TO NEW BRAIN IMAGING TECHNOLOGIES, EXCITING INSIGHTS ARE RAPIDLY PILING UP

Until quite recently, anatomists desiring a peek inside the human brain had to content themselves with dissecting dead tissue or, when the rare opportunity arose, examining people who had traumatic skull injuries. Only in the past few decades has it been possible to glimpse, in real time, the working brain in action. First came computerized tomography or CT scans, which use a series of X-rays to create a three-dimensional picture. Next, positron emission tomography, or PET scanning, revealed blood flow and metabolism by tracing the path of an injected radioactive chemical. Magnetic resonance imaging (MRI), now gaining the edge in brain research, produces detailed views of both anatomy and brain activity by agitating the body's hydrogen atoms using magnetic fields. Result: a sudden wealth of clues about everything from the mechanisms behind mental illnesses to what happens when a memory is triggered or a skill is learned.

"The details, shapes, and patterns in the brain are exquisite on aesthetic grounds alone. But then you realize this is the machinery of thought," marvels Carl Schoonover, a doctoral candidate in neuroscience at Columbia University who was so inspired by the imagery now possible that he created a coffee-table book, *Portraits of the Mind*, to give it center stage.

For all that these technologies promise to reveal, equally remarkable is the boost they already give physicians in practice. Victims of severe concussions are imaged quickly to pinpoint internal bleeding. Stroke patients are scanned to gauge whether clot-busting drugs will be effective. And surgeons can locate and map their way around territories vital to language, vision, and motor skills before plunging in the knife. Imaging is especially important for doctors working with the brain, says Joseph A. Helpern, professor and vice chairman for research in radiology at the Medical University of South Carolina in Charleston. "You don't want to have to cut into it to find out what's going on."

Even the lowly microscope, used to study tissue sliced out of an animal brain or a human brain after autopsy, has been making significant new contributions. The problem with microscopes that use light (remember your high school's equipment?) is that numerous structures, including many all-important synapses where neurons communicate, are smaller than a light wave and thus invisible. But treating the samples with dyes that can be switched on or off by pulses of light allows scientists to view ever-tinier structures. And microscopes relying on much smaller electron waves to magnify and illuminate brain tissue are powerful enough now to reveal even the minuscule spines, thought to store memory, that protrude from the tentacle-like dendrites of a single neuron.

One cutting-edge microscope can even view those spines in a living animal. New York University School of Medicine scientists recently used this "two-photon" microscope, which employs infrared light and fluorescent dyes, to probe the brains of mice before and after challenging them by, for example, changing something in their cages or speeding up their walkway. During the course of several weeks, the microscope revealed a correlation between these experiences and the creation of new spines. "We're documenting how daily sensory experiences leave minute but permanent marks in the brain," says study coauthor Guang Yang, assistant professor of anesthesiology.

It's the study of the living human brain, however, that has seen the most exhilarating leaps. In some cases, older technologies are offering up new secrets; PET scans, for instance, capture neurons reacting when a person is under

A PET scan can reveal blood flow by tracing an injected radioactive substance.

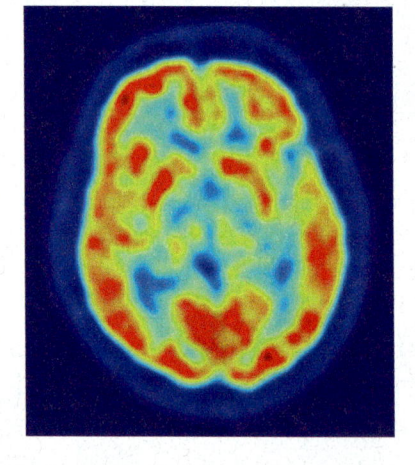

stress, and electrodes inserted into the brain reveal which neurons fire when a person thinks certain thoughts. But many of the latest findings hail from two advances in magnetic resonance imaging, "functional" and "diffusion" MRIs.

Back in the early 1990s, scientists realized that MRI magnets could be used to get images of where blood and nutrients rush when an area of the brain becomes engaged—when, say, a person is instructed to think a certain thought. These functional MRIs can then be compared to pictures taken minutes earlier. "I refer to functional MRI as a 'mindoscope,' because it allows us to connect the intangibles of conscious experience with the structure and function in the brain," says Joy Hirsch, who directs the Functional MRI Research Laboratory at Columbia University Medical Center in New York and helped curate "Brain: The Inside Story," an exhibit running through August 2011 at the city's American Museum of Natural History. While fMRIs are imperfect instruments, not least because blood takes a while to respond whereas neurons fire in a flash, the fact that subjects don't need to be cut into or injected with radioactive substances has made them the tool of choice for many researchers.

Diffusion MRI, meanwhile, relies on the understanding that water moving around inside a structure can reveal the structure's shape. For example, monitoring the water inside axons—those long, slender projections that carry nerve impulses but are too tiny to be detected with normal MRI—reveals each axon's position. Similarly, water leaking in unexpected places likely indicates defects in the axon's myelin coating, thought to be a factor with multiple sclerosis and other conditions. An even more advanced technique developed by Helpern and researcher Jens Jensen at New York University School of Medicine, diffusional kurtosis imaging, uses mathematical formulas to yield more detailed

Columbia's Joy Hirsch studies how the brain lights up when asked to perform various tasks.

Surgeons can locate and **MAP THEIR WAY** around territories vital to language, vision, and motor skills

information about the diffusing water. In preliminary findings not yet published, DKI scans correctly identified people with mild cognitive impairment, a condition that increases the odds of later Alzheimer's. "We're looking for changes in the brain tissue microarchitecture before the brain shows evidence of shrinkage, akin to an engineer seeking out fissures inside a building's wall well before the wall

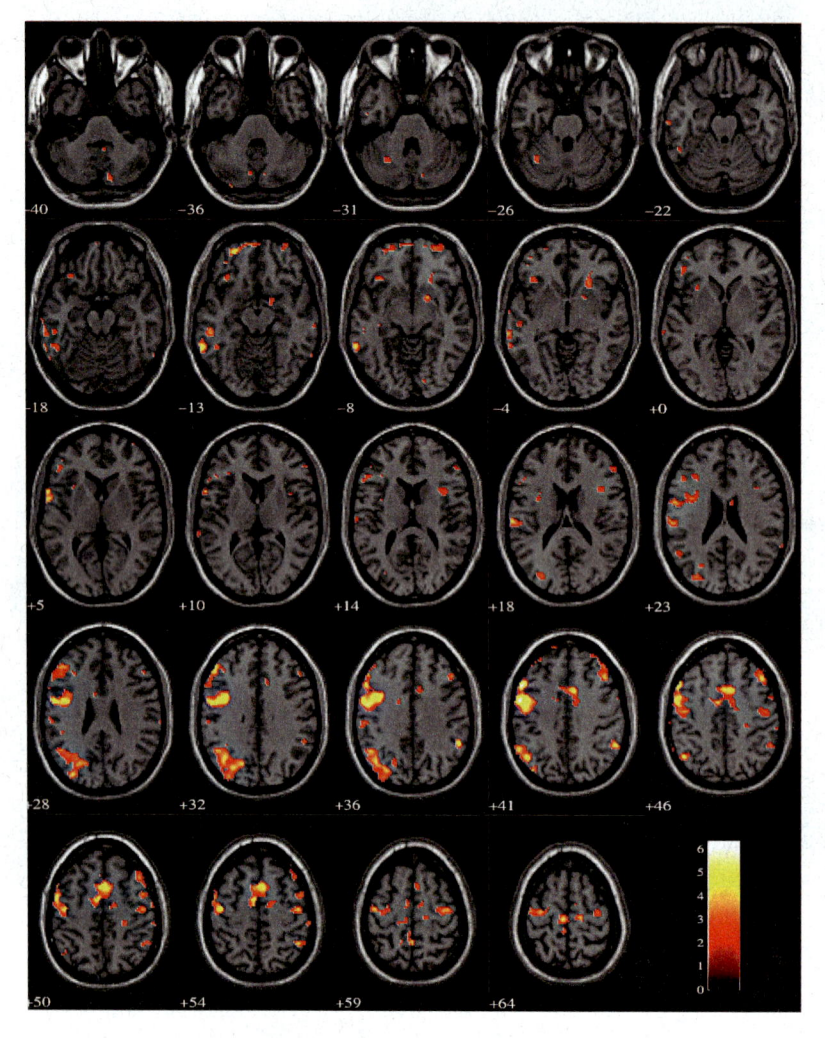

These fMRI scans show brain activity as a person reads and makes sense of a calendar.

less reasoned. Subsequent studies of people with ventromedial damage have found that, as would be expected, they don't have the same difficulty pushing the man off the bridge.

"It's good that we have strong emotional reaction to committing an act of violence. But where that violence would actually save lives, this automatic response may not be the best," Greene says. Reworking the footbridge scenario to remove some of the emotion—say, by creating a trapdoor activated by the press of a button—appears to quiet that part of the brain and make the act more tolerable. This is no esoteric exercise: Understanding this response could help people overcome their gut-level opposition to such emotionally charged practices as physician-assisted suicide or organ donation, Greene says.

Differences in the ADHD brain. "We know that during the teenage years the brain undergoes significant structural development and becomes much more architecturally complex," says Helpern. Using DKI to scan the prefrontal cortex of a dozen teens with attention deficit hyperactivity disorder and an equal number without, Helpern's team has just revealed strikingly different developmental progress. In the prefrontal cortex, both the white matter (where the myelin-sheathed axons aiding cell communication are located) and gray matter (home of the neuron cell bodies) developed less extensively in kids with ADHD. A difference in white matter has previously been a suspect in the decision-making challenge that is a hallmark of ADHD, but Helpern thinks this finding, published in the January issue of the *Journal of Magnetic Resonance Imaging,* is one of the first objective measures showing that the microstructure of the gray matter is different, too. The payoff of understanding the ADHD brain should eventually be more effective therapies.

begins to collapse," Helpern says. One day this may produce the holy grail of brain research, a way to identify and help people at risk for further impairment before the damage is done. Among brain researchers' other areas of exploration:

How emotion affects decision-making. Philosophers have long been stymied by the conflicting responses people give when faced with two seemingly similar moral dilemmas. Told that a runaway train is heading for five people on the track, most people find it acceptable to throw a switch and divert the train toward a single man in harm's way, yet they rebel at the idea of pushing the man off a footbridge onto the track to save the same five people. Using fMRI to image brain areas involved in decision-making, Joshua Greene, assistant professor of psychology at Harvard, has begun untangling the mystery. The prospect of actively shoving a man to his death seems to trigger an emotion-related part of the prefrontal cortex, the "ventromedial" section, while pondering the more neutral action of throwing a switch lights up areas involved with rational thought in the "dorsolateral" section nearby. Once the emotional system is engaged, choices become

The limits of willpower. Every dieter knows that keeping the weight off is often harder than losing it. By performing fMRI scans on six obese individuals as they viewed photographs of food, both before and after they lost 10 percent of their body weight, Columbia University researchers have illuminated at least part of the reason. When the participants were shown the pictures after they'd lost the weight, sections of the brainstem involved in processing rewards were much more active than they had been in the pre-diet scans, presumably making that chocolate cake tougher to resist than in the past. Then the subjects got a shot of the appetite-dampening hor-

UNDERSTANDING YOUR BRAIN

mone leptin, which unhelpfully becomes less plentiful after dieting. Sure enough, with the artificial help the reward center returned to its pre-dieting calm, when logically it would be easier to push the cake away. "We're realizing that if you want to study obesity, you've got to look in the brain," says Hirsch, coauthor of the study. "It's clear there's so much more than willpower going on in a dieter struggling with overeating."

Gender's relationship to mood. Why do depression and anxiety disorders strike women more frequently than men? Two preliminary studies suggest key brain differences. When scientists at Children's Hospital of Philadelphia stressed both male and female rats by forcing them to swim, then examined slices of their brains using an electron microscope, they found the males had an ingenious stress-reduction response: They tucked some receptors for a stress-inducing hormone inside their brain cells. Female rats, by contrast, kept all the receptors exposed on the surface. In addition, the hormone bound more tightly to the receptors in the female brains. "The females were more sensitive to the stress neural hormone, and they didn't adapt as readily," says researcher Rita Valentino, director of stress neurobiology at the hospital. Similarly, when PET scans were performed on 46 men and women at the Karolinska Institutet in Stockholm, Sweden, the women turned out to have significantly more binding sites in certain parts of their brains for a chemical believed to be associated with depression.

The neuron-thought connection. Epilepsy patients sometimes need to have electrodes implanted inside their brains to reveal the sites of their seizures. An international team of scientists has taken advantage of this access to better understand the biology of memory. In a highly publicized 2005 study of volunteers with the electrodes, the team found that a specific neuron became activated when a patient viewed photographs of a well-known celebrity but not when he or she looked at photos of other stars or similar-looking women. Thanks to the discovery of what's been called the "Jennifer Aniston cell," scientists have realized that neurons are much more specialized than was previously believed, says coauthor Christof Koch, professor of biology and engineering at California Institute of Technology. The knowledge may one day lead to ways to identify and repair damaged neurons that impair certain memories.

How the mind focuses. Ever wonder how you can concentrate on a book in a noisy, crowded bus? Joy Hirsch's team has discovered the mechanisms that the brain apparently uses to stay focused in distracting situations. While monitoring the temporal lobe via fMRI, the researchers asked people to categorize, by profession, photographs of familiar actors or politicians as they flashed before them. In some cases, the labels on the pictures conflicted with the images; a photo of an actor might be marked with a politician's name. To help the brain focus on the images, extra blood rushed to the area believed to process faces. When the subjects were instructed to ignore the photos and instead categorize the profession based on the written name, the extra energy went to the "reading" center. Understanding how these processes "are engaged to either enhance or regulate behavior opens new doors for possibilities for treating many disorders of decision-making, including, perhaps, addictions," Hirsch says.

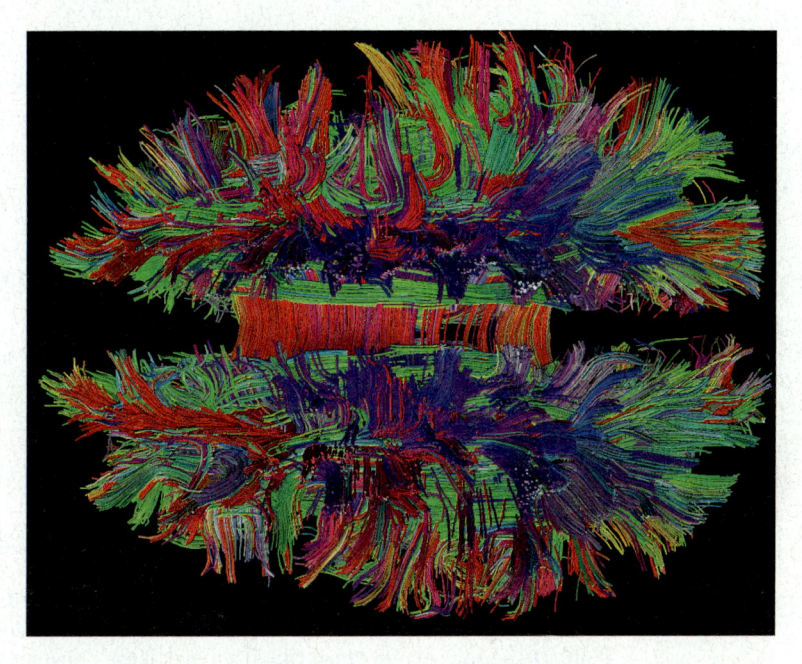

Integration of the whole network. Crowds simultaneously logging onto a single website can slow or crash the system. Neurologists believe the same sort of jam-up happens when something affects a part of the brain, be it a migraine or a devastating tumor. To better understand how the brain's wiring is interconnected, an international consortium of researchers is now creating a comprehensive diagram of the entire circuitry: the major highways leading to the smaller roads, down to each individual driveway or single neuron, says Olaf Sporns, professor of psychological and brain sciences at Indiana University. Using diffusion MRI on a handful of people, Sporns's team has already begun getting a sense of the routes within the cerebral cortex. The goal of this Human Connectome Project is to gather the wisdom that will come from imaging some 1,200 brains. ●

By Meryl Davids Landau

A colorful diffusion MRI scan relies on the water in the brain's axons to reveal their position.

PATRIC HAGMANN—UNIVERSITY HOSPITAL CENTER OF LAUSANNE, SWITZERLAND

Your Teen's Amazing Brain

Behold the American teenager, a lump in a hoodie who's capable of little more than playing "Grand Theft Auto," raiding the liquor cabinet, and denting the minivan, thanks to a brain so unformed that it's more like a kindergartner's than a grown-up's. That's the message that seemed to emerge from the past decade's neuroscientific discoveries: that the brain, once thought to be virtually complete by age 6, is very much a work in progress during adolescence and *not* to be trusted. But experts are realizing that the popular parental response—to coddle teens in an attempt to shield them from every harm—actually may be counterproductive.

Yes, teenagers make woefully errant decisions that factor big in the 13,000 adolescent deaths each year in the United States. And yes, their unfinished brains appear to be uniquely vulnerable to substance abuse and addiction. But they also are capable of feats of learning and daring marvelous enough to make a grown-up weep with jealousy. How they exercise these capabilities, it now appears, helps shape the brain wiring they'll have as adults. "You have this power you're given," says Wilkie Wilson, director of DukeLEARN, a program at Duke University designing a curriculum for high schools to use to teach teenagers (and their parents) how to best protect and deploy those brains. Far from coddling the kids, he says, Mom and Dad need to figure out how to allow enough "good" risk-taking to promote growth and prevent wasted talent—while also avoiding disaster.

It can be a nerve-racking exercise. "These kids are such a crazy mix of impulsiveness and shrewdness," says Marcia Harrington, a survey researcher in Silver Spring, Md., who recalls the time she thought her then 16-year-old daughter, Alexandra Plante, had sleepover plans, when in fact the girl had ditched school and flown to Chicago to visit an acquaintance she'd met briefly during a family trip. The scheme was revealed only because bad weather delayed the flight home. After that trip, Alex applied her daring spirit to becoming an emergency medical technician and volunteer for the local fire department, and she's now a junior in college.

Still pruning. It wasn't until the 1990s, when MRI scans became a common research tool, that scientists could peek into the teenage cranium and begin to understand why adolescents can be such risk-takers. What they found astonished them. The brain's gray matter, which forms the bulk of its structure and processing capacity, grows gradually throughout childhood, peaks around age 12, and then furiously "prunes" underused neurons. By scanning hundreds of children as they've grown up, neuroscientists at the National Institute of Mental Health have been able to show that the

UNDERSTANDING YOUR BRAIN

pruning starts at the back of the brain and moves forward during adolescence. Regions that control sensory and motor skills mature first, becoming more specialized and efficient. The prefrontal cortex, responsible for judgment and impulse control, matures last. Indeed, the prefrontal cortex isn't "done" until the early 20s—and sometimes even later in men.

Meantime, the brain's white matter, which acts as the cabling connecting brain parts, becomes thicker and better able to transmit signals quickly. Recent research shows that this myelination process of white matter continues well past adolescence, perhaps even into middle age.

Now, dozens of researchers are studying how all these changes might affect adolescent behavior and shape adult skills and behavior. The maturation lag between emotional and cognitive brain centers may help explain why teenagers get so easily upset when parents see no reason, for example; teens seem to process input differently than do adults. In one experiment, young teenagers trying to read the emotions on people's faces used parts of the brain designed to quickly recognize fear and alarm; adults used the more rational prefrontal cortex. University of Utah professor of psychiatry Deborah Yurgelun-Todd, the researcher who led this work, believes young teens are prone to read emotion into their interactions and miss content. Therefore, parents may have better luck communicating with middle schoolers if they avoid raising their voices and instead explain how they're feeling.

Other experiments shed light on why even book-smart teenagers come up short on judgment: Their brain parts aren't talking to one another properly. When Monique Ernst, a child psychiatrist and neurophysiologist at NIMH, uses functional MRI to watch teenage and adult brains engaged in playing a gambling game, she finds that the "reward" center lights up more in teens than in adults when players are winning, and the "avoidance" region is less activated in teens when they're losing. There's also less activity in teens' prefrontal cortex, which adults use to mediate the "yes" and "no" impulses from other brain regions. "The hypothesis is that there is this triumvirate of brain regions that needs to be in balance" in order to produce wise judgments, says Ernst, whether that's to wear a seat belt or use contraception.

Adult guidance. There is as yet no proven link between bright blobs on an MRI and real-life behavior, but researchers are trying to make that connection. Laurence Steinberg, a developmental psychologist at Temple University, found that teenagers in a simulated driving test were twice as likely to drive dangerously if they had two friends with them—and brain scans showed that the reward centers lit up more if teens were told that friends were watching.

A savvy parent might conclude that what's needed in the teen years is more guidance, not less. In fact, study after study has shown that one of the most powerful factors in preventing teenage pregnancy, crime, and drug and alcohol abuse is time with responsible adults. "It doesn't have to be parents," says Valerie Reyna, a professor of psychology at Cornell University. Reyna thinks adults also need to teach what she calls "gist" thinking, or the ability to quickly grasp the bottom line. Instead, she says, teenagers often overthink but miss the mark. When Reyna asks adults if they'd play

Coddling RISK-TAKING TEENS to shield them from every harm actually may be counterproductive

Russian roulette for $1 million, they almost universally say no. Half of teenagers say yes. "They'll tell you with a straight face that there's a whole lot of money, and they're probably not going to die. It's very logical on one level, but on another level, it's completely insane."

If it's any comfort, the evidence suggests that teenagers' loopy behavior and combativeness are hard-wired to push them out of the nest. Adolescent primates, rodents, and birds also hang out with their peers and fight with their parents, notes BJ Casey, a teen brain researcher who directs the Sackler Institute at Weill Medical College of Cornell University in New York City. "You need to take risks to leave your family and village and find a mate."

The revved-up adolescent brain is also built to learn, the new research shows, and those teen experiences are crucial. Neurons, like muscles, operate on a "use it or lose it" basis; a teenager who studies piano three hours a day will end up with different brain wiring than someone who spends that same time shooting hoops or playing video games. Only in early childhood, it turns out, are people as receptive to new information as they

DID YOU KNOW...

During the first five months in the womb, a baby's brain develops at the blistering pace of a half-million nerve cells, or neurons, each minute. By the time it is fully formed in the early 20s, the brain contains about 100 billion neurons, some of which span more than three feet in length.

SOURCE: "BRAIN: THE INSIDE STORY," AMERICAN MUSEUM OF NATURAL HISTORY

are in adolescence. The human brain is designed to pay attention to things that are new and different, a process called salience. Add in the fact that emotion and passion also heighten attention and tamp down fear, and teenagerhood turns out to be the perfect time to master new challenges. "You are the owners of a very special stage of your brain development," Frances Jensen, a neurologist at Children's Hospital Boston, tells teenagers in her "Teen Brain 101" lectures at local high schools. "You can do things now that will set you up later in life with an enhanced skill set. Don't waste this opportunity."

Jordan Dickey certainly seized opportunity. As a 14-year-old high school freshman, he asked his father for a $26,000 loan to start a business. The Dickeys, of Ramer, Tenn., raised a few head of cattle, and Jordan had noticed that people paid a lot more for hay in square bales than in less convenient round bales. After doing a feasibility study as an agriculture class project, Jordan persuaded his dad to give him a three-year loan to buy a rebaling machine. He worked nights and weekends, mowing, raking, and rebaling; paid friends $7 an hour to load the bales into a trailer; and hired drivers to deliver the hay to local feed marts, since he was too young to drive. "It taught me how to manage my own money," Jordan says.

That's an understatement. Not only did he pay off the loan in one year, he made an additional $40,000. Now 19, Jordan is paying for a big chunk of his education at Bethel University in Tennessee, where he's a sophomore.

Teens can apply the new findings to learn more without more study, notes Wilson, whose Duke-LEARN curriculum should be ready for a pilot in about a year. Key points:

● Brains need plenty of sleep because they consolidate memory during slumber.

● The brain's an energy hog and needs a consistent diet of healthful food to function well.

● Drugs and alcohol harm short- and long-term memory.

Teens' predisposition to learn has a bearing on the vexing issue of teenage drinking, smoking, and drug use. Neuroscientists have discovered that addiction uses the same molecular pathways that are used in learning, most notably those involving the neurotransmitter dopamine. Repeated substance use permanently reshapes those pathways, researchers say. In fact, they now look at addiction as a form of learning: Adolescent rats are far more likely to become hooked than adults.

And epidemiological studies in humans suggest that the earlier someone starts using, the more likely he or she is to end up with big problems. In 2008, a study tracking more than 1,000 people in New Zealand from age 3 to age 32 found that those who started drinking or using drugs regularly before age 15 were far more likely to fail in school, be convicted of a crime, or have substance abuse problems as an adult. "You can really screw up your brain at this point," says Jensen. "You're more vulnerable than you think."

The new brain science has been used as a weapon by both sides of the drinking-age debate, though there is no definitive evidence for a "safe" age. "To say that 21 is based on the science of brain development is simply untrue," says John McCardell, former president of Choose Responsibility, which advocates lowering the drinking age to 18. But there's also no scientific basis for choosing 18. The bottom line for now, most experts agree: Later is better.

Jay Giedd, an NIMH neuroscientist who pioneered the early MRI research on teen brains, is fond of saying that "what's important is the journey." Researchers caution that they can't prove links between brain parts and behavior, or that tackling adult-size challenges will turn teenagers into better adults. But common sense suggests that Nature had a reason to give adolescents strong bodies, impulsive natures, and curious, flexible minds. ●

By Nancy Shute

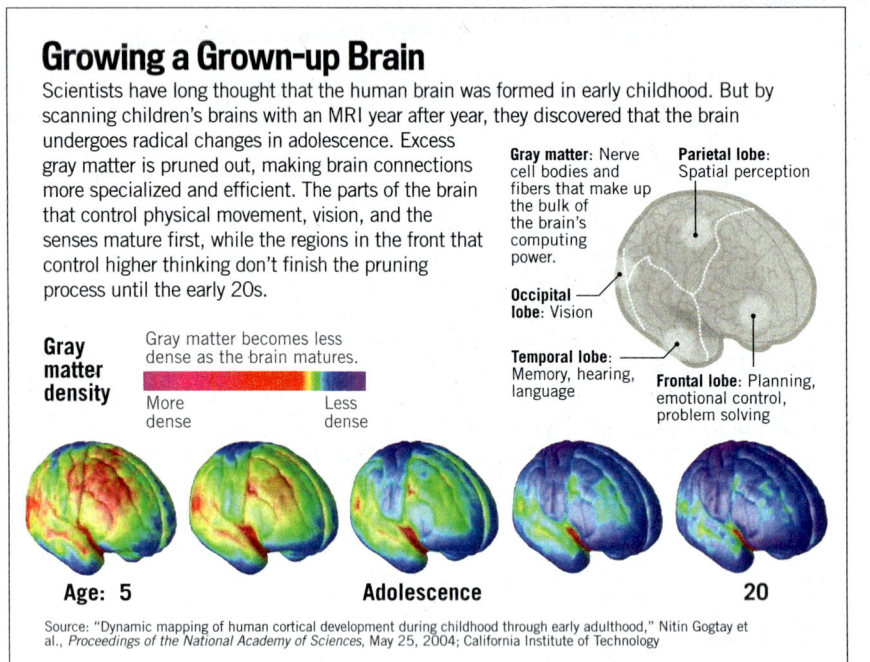

Growing a Grown-up Brain

Scientists have long thought that the human brain was formed in early childhood. But by scanning children's brains with an MRI year after year, they discovered that the brain undergoes radical changes in adolescence. Excess gray matter is pruned out, making brain connections more specialized and efficient. The parts of the brain that control physical movement, vision, and the senses mature first, while the regions in the front that control higher thinking don't finish the pruning process until the early 20s.

Gray matter: Nerve cell bodies and fibers that make up the bulk of the brain's computing power.

Occipital lobe: Vision

Temporal lobe: Memory, hearing, language

Parietal lobe: Spatial perception

Frontal lobe: Planning, emotional control, problem solving

Gray matter density — Gray matter becomes less dense as the brain matures. (More dense → Less dense)

Age: 5 — Adolescence — 20

Source: "Dynamic mapping of human cortical development during childhood through early adulthood," Nitin Gogtay et al., *Proceedings of the National Academy of Sciences*, May 25, 2004; California Institute of Technology

Rewiring Your Decision-Making

THE MENTAL SHORTCUTS PEOPLE RELY ON IN DAILY LIFE CAN OFTEN LEAD THEM INTO TROUBLE

Over time, humans have evolved cognitive rules of thumb, called "heuristics," to help them efficiently navigate the many choices confronting them each day. Heuristics save time and energy, notes award-winning health and science writer Wray Herbert in his new book, On Second Thought: Outsmarting Your Mind's Hard-Wired Habits, *but they can also distort a person's thinking, sometimes with deadly results.*

On Feb. 12, 1995, a party of three seasoned backcountry skiers set out for a day on the pristine slopes of Utah's Wasatch Mountain Range. Steve Carruthers, 37 years old, was the most experienced of the group, though they were all skilled skiers and mountaineers. Carruthers had been over these hills many times and was intimately familiar with the terrain. He and the two others planned to trek over the divide from Big Cottonwood Canyon to Porter Fork, the next canyon to the north.

As the skiers headed across a shallow, treed expanse, they triggered an avalanche. More than 100 metric tons of snow roared down the mountainside at 50 miles an hour, blanketing the slope and pinning Carruthers against an aspen. Another party heard the avalanche and rushed to the rescue, but by the time they dug Carruthers out, he was unconscious. He never regained awareness. The two other skiers in Carruthers's group survived, but they faced some serious criticism back home. What were they thinking? This pass was well known as avalanche terrain, and February was considered high hazard season. The chatter in the tight-knit skiing community was that the experienced Carruthers had ignored obvious signs of danger and tempted fate.

None of this rang true to Ian McCammon. He had known Carruthers for years, and the two had been climbing buddies at one time. Sure, Carruthers may have been a risk-taker when he was younger, but he had matured. Just recently, while the two men were riding a local ski lift together, Carruthers had talked adoringly about his wife, Nancy, and his 4-year-old daughter, Lucia. His days of derring-do were over, he had told McCammon. It was time to settle down.

So what happened on that fateful afternoon? What skewed this experienced backcountry skier's judgment so that he would put himself and his party in harm's way? Did he perish in an avoidable accident? Saddened and perplexed by his friend's death, McCammon determined to figure out what went wrong. An experienced backcountry skier in his own right and a wilderness instructor, he is also a scientist with a Ph.D. in mechanical engineering. As a researcher at the University of Utah, he worked on robotics and aerospace systems for NASA and the Defense Department.

McCammon already knew snow science pretty

HEURISTICS CAN BE TRAPS, as they were in the frozen mountain pass where Steve Carruthers perished

well, so he began reading everything he could on the science of risk and decision-making. He ended up studying the details of more than 700 deadly avalanches that took place between 1972 and 2003, to see if he could find any commonalities that might explain his friend's untimely death. McCammon systematically categorized all the avalanches according to several factors well known to backcountry skiers as risks: recent snowfall or windstorm, terrain features like cliffs and gullies, thawing and other signs of instability, and so forth. He computed an "exposure score" to rate the risk that preceded every accident.

Then he gathered as much information as he could on the ill-fated skiers themselves, all 1,355 of them: the makeup and dynamics of the skiing party, the expertise of the group leader as well as the others, plus anything that was known about the hours and minutes leading up to the fatal moment. Then he crunched the data. His published results were intriguing. He found many patterns in the accidents, including several poor choices that should not have been made by experienced skiers. He concluded that these foolish decisions could be explained by several common thinking lapses, and he wrote up the work in a paper titled "Evidence of Heuristic Traps in Recreational Avalanche Accidents." The paper has become a staple of modern backcountry training and has no doubt saved many lives.

Heuristics are cognitive rules of thumb, hardwired mental shortcuts that people use each day to make routine decisions and judgments. The study of heuristics is one of the most robust areas of scientific research today, producing hundreds of articles a year, yet the concept is little known outside the labs and offices of academia.

Heuristics are normally helpful—indeed, they are crucial to getting through the myriad decisions we face every day without overthinking every choice. But they're imperfect and often irrational. They can be traps, as they were in the frozen mountain pass where Carruthers perished.

Much has been written in the past couple of years about the wonders of the rapid, automatic human mind and gut-level decision-making. And indeed the unconscious mind is a wonder. But it is also perilous. The shortcuts that allow us to navigate each day with ease are the same ones that can potentially trip us up in our ordinary judgments and choices, in everything from health to finance to romance.

Most of us are not backcountry skiers, and we will probably never face the exact choices that Carruthers and his friends confronted at Gobbler's Knob. But just because the traps are not life-threatening does not mean they aren't life-changing. Here are a few of the cognitive rules of thumb that shaped the backcountry skiers' poor choices—and may be shaping yours in ways you don't even recognize.

The familiarity heuristic. This is one of the mental shortcuts that McCammon identified as a contributing factor in many of the avalanche incidents he studied. It was one of the original heuristics identified and studied by pioneers in cognitive science. Basically, it tells us: If something comes quickly to mind, trust it. To the mind, familiar equals better and safer.

Heuristics are amazing time-savers, which makes them essential to our busy lives. Many, like the familiarity heuristic, are an amalgam of habit and experience. We don't want to think through every minor choice we make every day, and we don't need to. But there are always risks when we stop deliberating. McCammon's avalanche victims, for example, were almost all experienced

---

DID YOU KNOW...

Synesthesia, a rare neurological condition, causes one sensory system to affect another, allowing us to "taste" words or "see" music. Russian composer Alexander Scriabin wrote music in color because of the hues he associated with different keys.

SOURCE: "BRAIN: THE INSIDE STORY," AMERICAN MUSEUM OF NATURAL HISTORY

UNDERSTANDING YOUR BRAIN

backcountry skiers, and indeed almost half had had some formal training in avalanche awareness.

This expertise didn't guarantee that they would make the smartest choices. Paradoxically, in fact, it may have hurt them. They were so familiar with the terrain that it seemed safe—simply because it always had been safe before. It was familiar, and thus unthreatening. The skiers let down their guard because they all remembered successful outings that looked pretty much the same as the treacherous one. In fact, McCammon found in his research that there were significantly more avalanche accidents when the skiers knew the specific locale, compared to parties exploring novel terrain.

The default heuristic. So familiarity and comfort can be traps. But the fact is, Carruthers's decision-making began to go wrong long before he had even commenced to wax his skis. It started back in the warmth of the living room, when he or one of his buddies said, "Hey, let's take a run out to Gobbler's Knob tomorrow."

At that point, another powerful cognitive tool was triggered, known as the "default" or "consistency" heuristic. With their adventure still an abstract notion, Carruthers's group no doubt discussed the conditions, the pros and cons, and made a deliberate assessment of the risks. But once they made their choice, the cold calculation stopped. They made a mental commitment to move forward.

Human beings have a powerful bias for sticking with a decision once made. We rely on this stay-the-course impulse all the time, often with good results. Constant switching can be perilous, in everything from financial matters to romantic judgments, so we have become averse to hopping around.

The skiers stuck to their plan because they were cognitively biased toward going ahead rather than switching gears. Most people do this hundreds of times a day, simply because it takes effort to switch plans. We stay in relationships that are going nowhere simply because it's easier than getting out. We buy the same brand of car our father did or hesitate to rearrange our stock portfolio for the same reason.

The acceptance heuristic. The skiers likely got additional mental nudging from what McCammon calls the "acceptance" or "mimicry" heuristic, which reflects our strong tendency to make choices that we believe will get us noticed—and more important, approved—by others. It's deep-wired, likely derived from our ancient need for belonging and safety. It can be seen in the satisfaction we get from clubs and other social rituals, like precision military formations and choral singing. It's a crucial element in group cohesion, but we often apply it in social situations where it's inappropriate—or even harmful, as it was in many of the accidents that McCammon studied. His analysis showed

Human beings have a powerful bias for sticking with a decision and not CHANGING COURSE

a much higher rate of risky decision-making in groups of six or more skiers, where there was a larger "audience" to please.

These are just a few examples. Some psychologists estimate that there are hundreds of powerful heuristics at work in the human brain, some working in tandem with others, sometimes reinforcing and sometimes undermining one another. The best way to rein in faulty reasoning is to recognize it. Once we do, then we have the power to engage the more deliberate part of our brain to make the best choices for each situation. ●

Excerpted from *On Second Thought: Outsmarting Your Mind's Hard-Wired Habits.* Copyright © 2010 by Wray Herbert. Published by Crown, a division of Random House Inc.

PHOTO ILLUSTRATION BY PETE MCARTHUR

UNDERSTANDING YOUR BRAIN

How Optical Illusions Deceive

YOUR BRAIN HAS SURPRISING LIMITATIONS
WHEN INTERPRETING WHAT YOU SEE

Consider one of the most basic magic tricks. A magician shows you a coin in the palm of his left hand and slowly closes his fist around it. When he reopens his fist, the coin is gone. He then shows you his right hand. Somehow, the coin is now there. The trick couldn't be simpler, but a skilled magician, understanding how easily the eye can be fooled, can still pull it off.

Over the past three decades, neuroscientists have made great progress in understanding how the human brain and eyes can be deceived by visual illusions. But in many cases, the new knowledge

Since the human visual system is FAIRLY LOW-RESOLUTION, it focuses well only on a small area

just confirms what practitioners, from artists to military strategists, have learned through trial and error over many centuries.

In *Sleights of Mind: What the Neuroscience of Magic Reveals About Our Everyday Deceptions*, authors Stephen Macknik and Susana Martinez-Conde explain that human eyes see the world in a resolution roughly equivalent only to that of a one-megapixel digital camera of limited scope. Your cellphone camera probably has better capabilities. The brain processes images two-dimensionally, but through various adaptations it translates these crude images into the vivid three-dimensional world we experience. That process involves shortcuts, though, and those shortcuts are what magicians, artists, and others can exploit to trick us.

Magicians, for example, "don't use the scientific method per se, but they try things and will make variations on their art based on what works and what doesn't," says Martinez-Conde, director of the Laboratory of Visual Neuroscience at the Barrow Neurological Institute in Phoenix. Among the most important keys to visual illusion is what she and Macknik (director of Barrow's behavioral neurophysiology lab) call "the spotlight of attention."

Since the human visual system is fairly low-resolution, it focuses well only on a small area, like the circle of a spotlight. Thus, a skillful illusionist can simply use a shiny coin to direct the viewer's "spotlight" to a visual field of his choosing. He is then able to slip a coin—out of that spotlight—into his other hand.

Innovators in a variety of disciplines continue to draw on a mixture of neuroscience and practical experience to find new ways to dazzle or confuse the eye and brain. Here are just a few examples:

Art and Design
Like magicians, artists have long been ahead of scientists in grasping how the brain processes images—and how to exploit that knowledge. The

Italian polymath and painter Leonardo da Vinci, for example, created the celebrated 16th-century portrait of Mona Lisa, whose enigmatic expression has prompted centuries of debate.

When you look directly at the woman's mouth, she seems to be smiling slightly. But when you stare directly at her eyes, the impression of her mouth changes. She seems to be smiling much more. Your changing perception depends on where you are looking and occurs because her smile is blurred. Leonardo used a technique called sfumato, delicate shadings that mute features and make subtle transitions from dark to light areas. Our higher-acuity central vision is specialized for seeing small detailed objects, and our lower-acuity peripheral vision is better at seeing larger, blurry images. (This is why you need to move your eyes when you read.)

For these reasons, Mona Lisa's smile is more apparent to our peripheral vision than to our central vision, says Harvard Medical School neurobiologist Margaret Livingstone, who has studied the painting. The result, Livingstone explains, is that

Pop artist Roy Lichtenstein's *House I* seems to change from two dimensions to three as you walk around it.

PHOTOGRAPHY BY BRETT ZIEGLER FOR USN&WR

Trompe l'oeil (French for **"FOOL THE EYE"**) is a technique that makes flat images look three-dimensional

when people move their eyes around the portrait, Mona Lisa's expression seems to change, making her seem uncannily lifelike.

Often, illusion has been used in art simply to show off the artist's skill and delight the viewer. This is possible because photoreceptors in your eyes convert light into electrochemical patterns that are processed in the retina. The information then travels up the optic nerve into the brain to be further processed. Yet while the brain draws optical information from the real world, the images it assembles from this data do not represent perfect copies. The brain can be misled because it must make assumptions to "fill out" the two-dimensional information it receives. You know a tree trunk is round not because your eyes tell you so, but because experience has filled out your understanding of everyday objects over time.

Trompe l'oeil (French for "fool the eye") was a crowd-pleasing technique developed by ancient Greek artists to make flat images look like vivid three-dimensional objects. This method involves the use of shading and vanishing points (that is, drawing parallel lines that recede to a single point in the distance) to create works that seem astonishingly real. Contemporary artists have taken trompe

UNDERSTANDING YOUR BRAIN

(Clockwise from left) Julian Beever with an example of his 3-D street art in New York; the Roman-designed Library of Celsus in Turkey; Leonardo da Vinci's *Mona Lisa*.

l'oeil to new heights. In cities around the world, passersby can walk across striking 3-D scenes, from a waterfall pouring down the center of a street to Spiderman climbing the face of a building.

Illusion has also been used to make architecture look even more impressive. In Ephesus, Turkey, the remains of the ancient Library of Celsus stand in a narrow lot. The library's architects strove to make the structure, built by the Romans around A.D. 135, seem of greater proportions than it actually was. To accomplish this, they made the rafters and capitals on the middle columns bigger than those on the outer ones. This effect creates a sense of wider space between the columns. Livingstone says this "perspective illusion" works because the columns in the top row are narrower than those on the bottom, though they look very similar. Your eye processes the upper columns as farther away, creating a greater sense of depth.

The designers added other clever touches, Livingstone notes. The edges of the podium were carefully sloped and the central doorway was made larger and taller than the others. These design choices further reinforced the overall impression of the library as being larger.

One of the most famous visual illusions in the world is found in the 17th-century Church of St. Ignatius of Loyola at Campus Martius in Rome. The church's interior was painted by Andrea Pozzo around 1685. If a visitor stands in a particular spot in the nave and looks up, he sees not the flat ceiling, but a soaring false dome, rising toward the sky, with arches and windows apparently letting light in. Pozzo employed quadratura, a technique that

CLOCKWISE FROM LEFT: STUART RAMSON—AP; GETTY IMAGES; AP

SPECIAL EDITION 29

U.S.News & WORLD REPORT
UNIVERSITY DIRECTORY

FINDING AMERICA'S BEST
COLLEGES
JUST GOT EASIER

Explore **top schools**. Discover in-demand degree programs. Take charge of your future at USNewsUniversityDirectory.com!

- Search our extensive directory of on-campus and online programs.
- Contact over 1,900 of America's best colleges and universities.
- Get up-to-the-minute tips from award-winning *U.S. News* reporters.
- Take advantage of $116 billion in grants, loans and scholarships with our free financial aid guide!

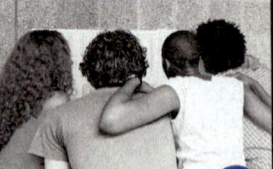

FREE FINANCIAL AID GUIDE!

College Funding 101: Demystifying the Financial Aid Process

| Bachelor's | Master's/MBA | Doctorate | Certificates |

FIND *YOUR* PERFECT PROGRAM: Business   **FIND NOW**

USNewsUniversityDirectory.com/Explore

Connect with us on facebook twitter

©2010 U.S. News University Connection, LLC. All rights reserved. MCID: 50405

used an advanced understanding of perspective to create the illusion that the actual architecture of a building was continuing upward and expanding. Pozzo also made liberal use of foreshortening, or compressing figures so that objects look more distant while adding depth so they appear three-dimensional. Alberto Pérez-Gómez, a professor of architecture at McGill University in Montreal, says Pozzo carefully created a specific vantage point where he intended the viewer to stand and gaze upward to maximize the effect.

More recently, artists have found new ways to play with 3-D perception of two-dimensional images. In his famous lithograph, *Ascending and Descending,* 20th-century Dutch artist M.C. Escher showed figures walking on a continuously rising rooftop staircase—a structure that could not be re-created in the real world. According to Livingstone, the brain interprets the visual universe using an enormous number of small parallel processors. In this case, each processor transmits tiny bits of information about discrete sections of the structure, including their distance and depth. Because the information from the local processors seems to make sense (sections of the stairway rising or falling), you accept the global image your brain assembles, even though it could not, in fact, exist.

Confusing the Enemy

Military deception is as old as warfare. In the ancient Roman navy, sailors dyed their sails and tunics a light blue so as to mimic the color of the sea. In the 19th century, British troops in India discarded their brightly colored uniforms (which made them inviting targets) in favor of bland khaki to blend in better with the terrain.

The military has capitalized on different visual weaknesses to deceive opponents. The brain has limited ability, for example, to judge the distance or dimensions of objects without known figures of fixed size and location to judge them against. On June 6, 1944, British planes made a nighttime drop of 500 paratrooper dummies, 3 feet tall, near Caen, France. Their goal was to delude the Germans into believing that the Allied invasion would start there and not on the beaches of Normandy. The scheme worked. Against the night sky, the dummies looked real, and the Germans diverted troops to the area.

The U.S. armed forces would adopt woodland patterns for uniforms and equipment as camouflage during the Vietnam War. But only in the 1980s did researchers seriously draw on advances in neuroscience to take combat uniforms to a new level. The result was the creation of "digital camouflage," which uses tiny multi-colored squares to create patterns on the outerwear of U.S. soldiers and Marines. According to Timothy O'Neill, a retired Army officer, former West Point engineering psychology professor, and Army consultant who helped pioneer the digital camo, the brain's visual systems can be divided into two parts, what he calls the "what is it" and "where is it" systems. The best camouflage tries to confuse both. The squares are arranged in precise patterns to fool various visual systems in the eyes and brain. Some elements of the patterns are designed to blend into a variety of backgrounds, while others break up the geometry of the wearer, preventing the enemy from identifying both the shape of a person or movements that are recognizably human.

This sort of visual camouflage may soon become outdated, however. In a quest that seems torn from science fiction (or Harry Potter), the Pentagon is in the early stages of developing its own "invisibility cloak." The garment would draw on physics, using materials that would steer light around the wearer. The ultimate optical illusion, of course, is duplicating the magician's trick of making someone disappear completely. ●

By Joshua Kucera

Digital camouflage disrupts the visual systems of potential foes, preventing them from identifying a human form and movement.

UNDERSTANDING YOUR BRAIN

Straight-A Students, Take Note

YOUR EMOTIONAL INTELLIGENCE MAY MEAN MORE TO YOUR SUCCESS IN LIFE THAN BOOK SMARTS

In his bestselling books *Emotional Intelligence* and *Social Intelligence*, psychologist **Daniel Goleman** explains why so many "brilliant" students who ace the SAT and achieve a perfect grade-point average flunk life: Their cognitive capabilities aren't the only ones that count. Equally key to success, it turns out, are self-awareness and an ability to understand and communicate feelings and to draw on them to make sound decisions. Goleman recently talked with writer Deborah Kotz about the latest findings on emotional intelligence.

Can we learn to be emotionally intelligent? Or is it, like IQ, something that we're born with?

Personality, in terms of your temperament, is largely determined by genetics, but emotional intelligence is mostly based on skills you acquire in life. It's about how you handle your emotions, how you handle relationships—your ability to feel empathy as well as how much self-mastery or self-discipline you have. Some people may be very adept socially but just can't get it together to finish their work because they lack self-management skills. They're lacking in emotional intelligence.

Emotional intelligence is about your ability TO FEEL EMPATHY and how much self-discipline you have

What we've learned in recent years is that the brain is neuroplastic; it generates 10,000 new stem cells every day. These "baby" cells migrate to where they're needed, to where we're learning, whether it's a new language or new set of social skills.

What can we do to raise our EQ levels?

Putting a high priority on face-to-face interactions can certainly help. Our increasing reliance on E-mail and social networking sites has hindered us from having full emotional interactions. When we communicate electronically, we lose all of the hundreds of emotional signals that go along with in-person interactions, like voice inflections, facial expressions, eye contact, and posture. Studies have shown there's a negativity bias in E-mail—that most of us interpret neutral E-mails as being critical in tone. This can hamper our relationships.

Electronic interactions also prevent input to the brain that enables us to mimic the actions of others: smiling, say, when our coworker smiles. Recently discovered nerve cells in the brain, called mirror neurons, have been shown to fire when one observes an emotional reaction in another, which causes that same emotion to be felt by the observer; it's why we say emotions are contagious. These mirror neurons are thought to evoke empathy, an essential component of emotional intelligence. And, no, you can't replace a personal encounter with an E-mail emoticon; your mirror neurons can't be that easily fooled.

The other problem with E-mail is that often it leads to cyber-disinhibition, which makes us far more likely to take risks and say exactly how we feel. Too many of us fire off missives before we have a chance to cool down. When you're expressing anger at a person face to face, you're more likely to hold back because your social brain, taking cues from the person standing before you, tells you there will be consequences if you express everything on your mind.

Gaining mastery over negative emotions like anger or sadness can be really tough. Is there a way to train the brain to deal better with adversity?

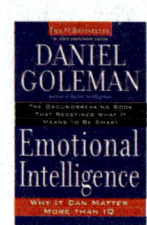

Our brain has circuitry that helps us manage distressing emotions; this circuitry can be enhanced, like a muscle, through training techniques. What happens when we get upset is that a part of our brain called the amygdala takes over; this emotional center literally hijacks the prefrontal cortex, which governs our decisions. You have the classic fight-or-flight response, during which you can't learn, innovate, or be flexible; your memory is hampered; and all you can feel is a threat. Some people feel this way much of the day, especially in this era of layoffs and foreclosures.

What the latest research has found, though, is that there's a difference between the left and right hemispheres of the brain when it comes to dealing with emotions. While the right prefrontal cortex can get hijacked by the amygdala, the left prefrontal cortex can just say "no" to the hijack and help to turn off the stress response. The old wisdom of counting to 10 when you're angry can help activate the left prefrontal cortex. And relationship expert John Gottman was right on the money when he said couples should take a 20-minute break when they're in the middle of a fight. We now know that more optimistic folks tend to have slightly more dominant left prefrontal cortexes, whereas hostile pessimists tend to have more dominant right prefrontal cortexes.

Can we convert to a "glass half full" person?

It's tough but not impossible. To cultivate your left prefrontal cortex, you need to take time off to rest and restore every day, whether through exercise, meditation, or a bubble bath. I favor mindfulness meditation, which entails simply observing what's happening around you, focusing on light, sounds, colors, and smells without any judgment or distraction. It's been shown in imaging studies to alter the brain. One landmark study by Richard Davidson at the University of Wisconsin–Madison found that stressed-out executives who practiced mindfulness meditation for 20 minutes a day for eight weeks were able to shift from a more activated right prefrontal cortex to a more activated left and recall what they loved most about their jobs instead of feeling burned out and overwhelmed.

How does social intelligence differ from emotional intelligence?

Social intelligence, which is a part of emotional intelligence, refers to our ability to understand and manage relationships. It includes empathy and social skill. Emotional intelligence, in my model, includes self-awareness and self-regulation of emotions, and then social intelligence.

Are you concerned that today's children are going to have a lower EQ than their parents?

Yes, especially with all the texting and E-mailing they do, even when they're with friends. I'm a big proponent of teaching kids to be emotionally intelligent, because it's going to matter so much later in life. If you simply tell children they're great, no matter what, they see through that and don't put much value in your compliments. Parents should praise their kids based on their abilities and achievements; self-awareness improves social interactions. Kids should also be encouraged to cultivate empathy and compassion; one way, for example, is by having face-to-face interactions and making eye contact during conversations. And, yes, parents should set limits on screen time. ●

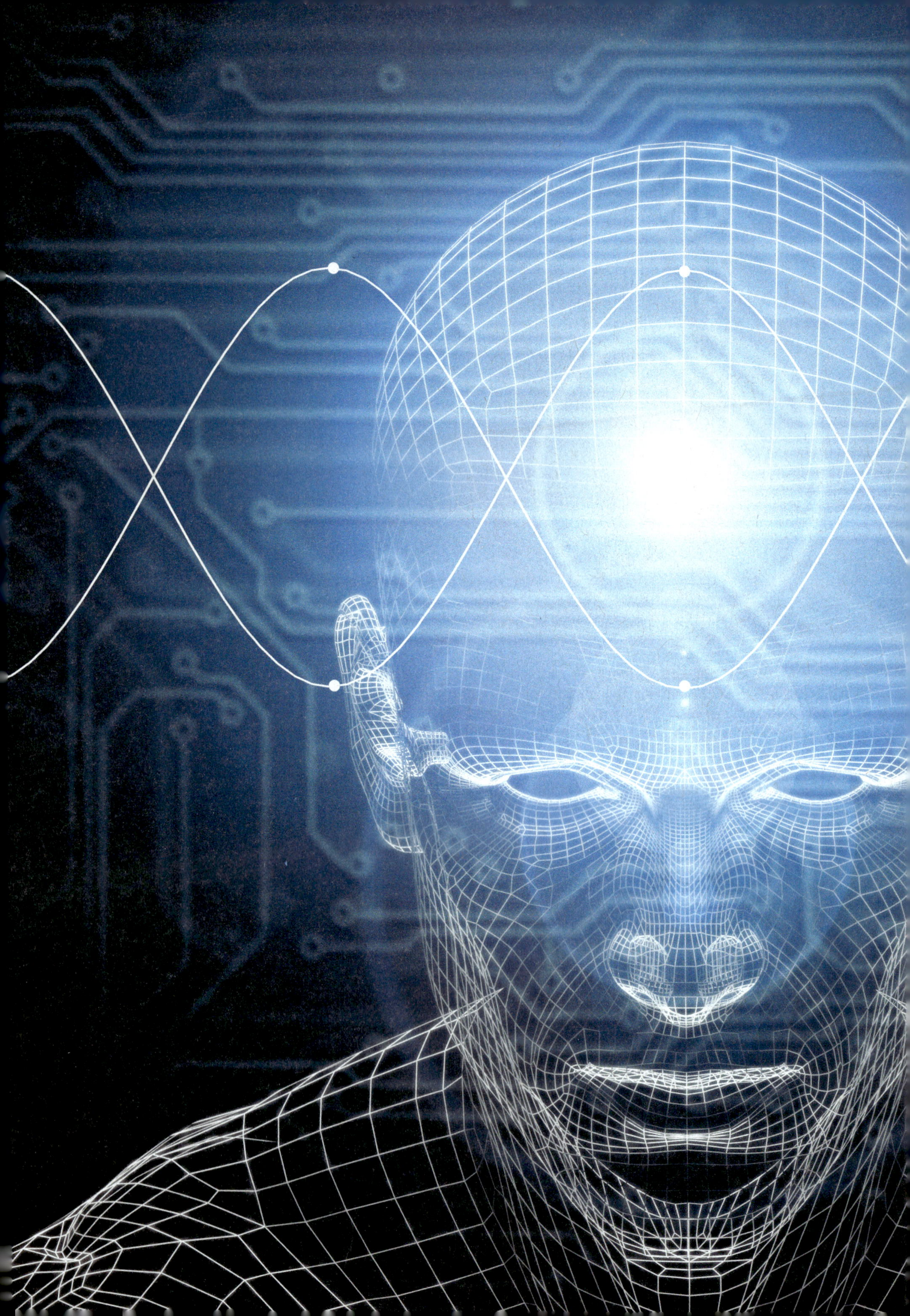

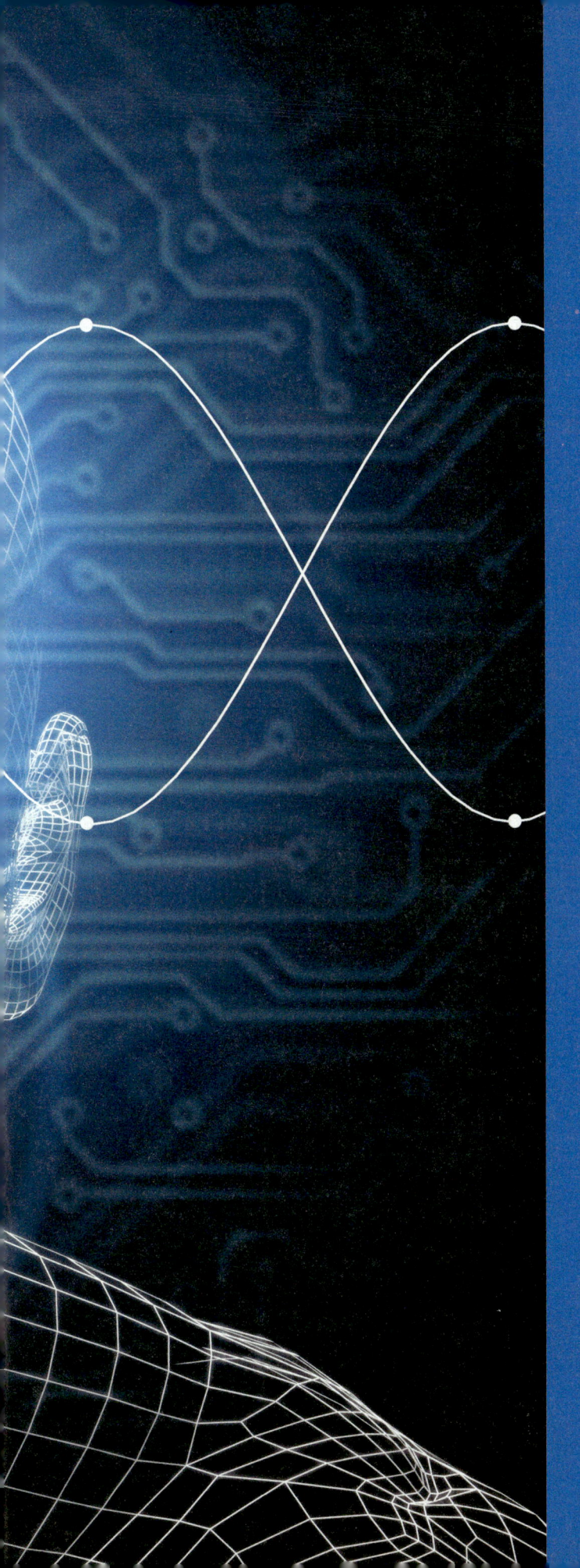

2.
Mysteries of the Mind

What goes on in the brain when people dream? **36**

Investigating the many benefits of sleep **41**

The science behind those transcendent out-of-body experiences **42**

Probing the puzzle of the psychopathic mind **45**

The brain chemistry behind that intense feeling of falling (and being) in love **48**

MYSTERIES OF THE MIND

What Dreams Are Made Of

Our eyes close and our mind begins to play: We are flying, we are falling; we argue again, but in a different place; we walk out on stage and find we've forgotten all of our lines; we share a heart-to-heart with a friend we haven't seen in years. Shakespeare wrote, "We are such stuff as dreams are made on," and 300 years later, Sigmund Freud gave the poetry a neat psychoanalytic spin when he called dreams "the royal road to the unconscious." The movies that unfold in our heads some nights are so powerfully resonant they haunt us for days—or inspire us. Mary Shelley dreamed of Frankenstein's monster before she created him on paper; the melody to "Yesterday" came to Paul McCartney as he slept.

Everybody dreams. Yet no one, throughout history, has fully grasped what the dreaming mind is doing. Are the nightly narratives a message from the unconscious to the conscious mind, as Freud believed? Or are they simply the product of random electrical flashes in the brain? Today, researchers aided by powerful technologies that reveal the brain in action are concluding that both schools of thought hold truth. "This is the greatest adventure of all time," says J. Allan Hobson, dream researcher and Harvard psychiatry professor emeritus. "The development of brain imaging is the equivalent of Galileo's invention of the telescope, only we are now exploring inner space instead of outer space."

Mind-brain dance. The dream researchers' latest tools, functional magnetic resonance imaging and positron emission tomography scanning, have been used for some time to capture the waking brain at work—making decisions, feeling frightened or joyous, coping with uncertainty. And those efforts have shown clearly that psychology and physiology are intimately related. In someone suffering from an anxiety disorder, for example, the fear center of the brain—the amygdala—lights up as neurons fire in response to images that trigger anxiety; it flickers in a minuet with the center of memory, the hippocampus. Scans of people who are sleeping, too, suggest that the same sort of mind-brain dance continues 24 hours a day.

"Psychology has built its model of the mind strictly out of waking behavior," says Rosalind Cartwright, professor emeritus of neuroscience at Rush University Medical Center's Graduate College in Chicago, who has studied dreams for most of her 88 years. "We know that the mind does not turn off during sleep; it goes into a different stage." Brain cells fire, and the mind spins. Problems find solutions; emotional angst seems to be soothed; out-of-the-box ideas germinate and take root.

Freud saw dreams as deeply buried wishes disguised by symbols, a way to gratify desires unacceptable to the conscious mind. His ideas endured for years, until scientists started systematically studying dream content and decided that actually,

> **Does dreaming help people learn? SOME SORT OF BOOST appears to happen during sleep**

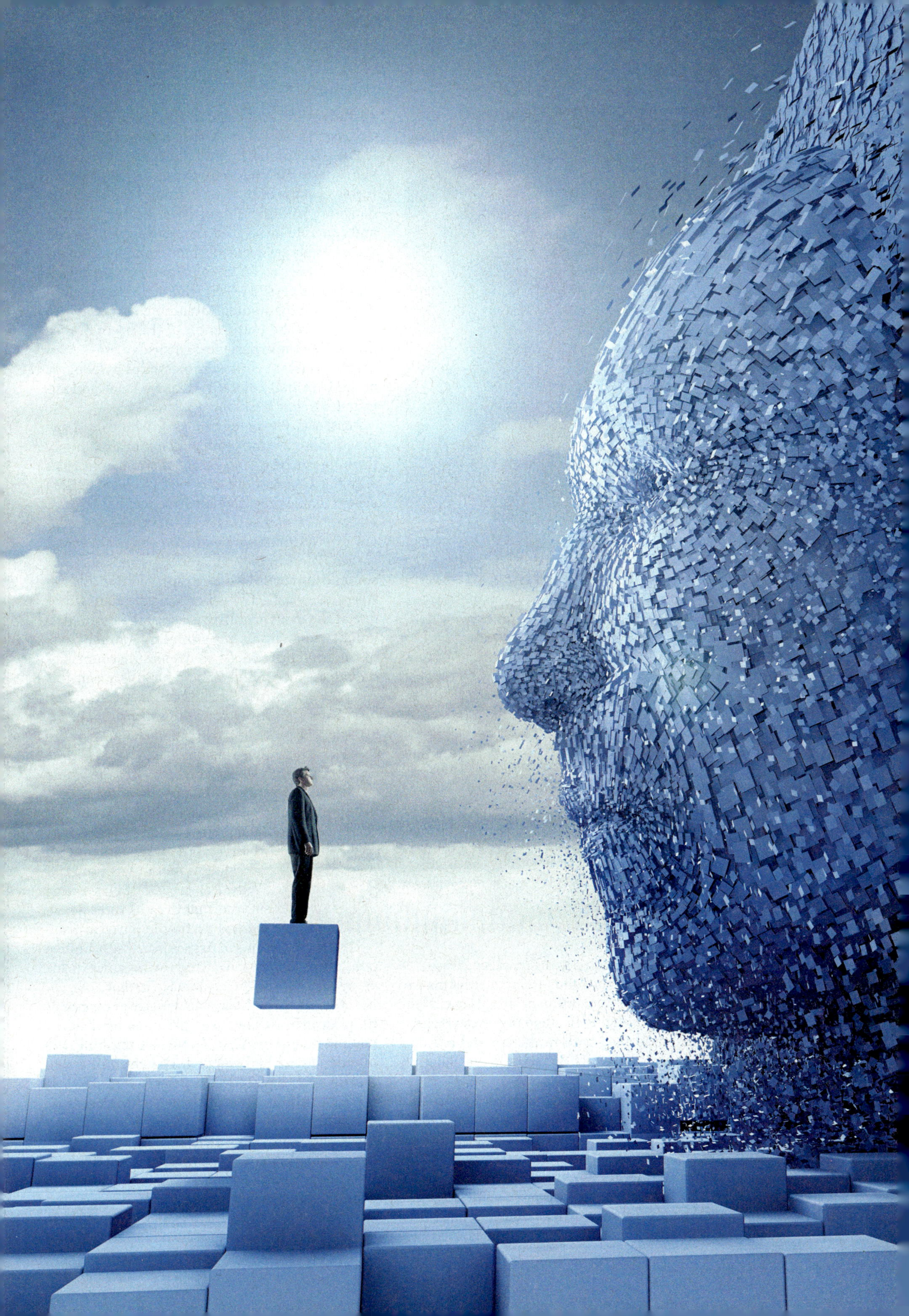

narratives and are held at Santa Cruz. (The narratives can be read at *www.dreambank.net*.)

Post-Freudians might argue that the monsters lurking in children's dreams signal a growing awareness of the world around them and its dangers. Young children describe very simple and concrete images, while the dreams of 9- and 10-year-olds get decidedly more complex. A monster that goes so far as to chase or attack might represent a person who is frightening to the child during waking hours. "Dreaming serves a vital function in the maturation of the brain and in processing the experiences of the day," says Alan Siegel, assistant clinical professor of psychology at the University of California–Berkeley and author of *Dream Wisdom*.

Nonsense. Physiology purists, who would say that dreams are the product of our brains simply flashing random images, got their start in 1953 with the discovery of rapid eye movement (REM) sleep. Using primitive electroencephalograms, researchers watched as every 90 minutes, sleepers' eyes darted back and forth, and brain waves surged. Then, in 1977, Harvard psychiatrists Hobson and Robert McCarley reported that during sleep, electrical activity picked up dramatically in one of the most primitive areas of the brain—the pons—which, by simply stimulating other parts of the brain, produced weird and disconnected narratives. Much like people looking for meaning in an inkblot, they concluded, dreams are the brain's vain attempt to impose coherence where there is none.

Or maybe that's not the whole story, either, said a young neuropsychologist at the Royal London School of Medicine 20 years later, when his findings hinted that dreaming is both a mental and a physical process. Mark Solms showed that dreams can't be explained as simple physical reactions to flashes from the primitive pons, since some of the most active dreamers in his study had suffered brain damage in that area. On the other hand, in those with damage to regions of the brain associated with higher-order motivation, passionate emotions, and abstract thinking, the nightly movies had stopped. That seemed a sign that dreams might indeed express the mind's ideas and motivations. "It is a mistake to think that we can study the brain using the same concepts we use for the liver," says Solms.

"From my perspective, dreaming is just thinking in a very different biochemical state," says Deirdre Barrett, who teaches psychology at Harvard and is author of *The Committee of Sleep*. The threads can be "just as complex as waking thought and

Frankenstein and Mary Shelley

Can man create life? A talk on evolution that considered the possibility so disturbed Mary Godwin that she went to bed and dreamed up Frankenstein's monster. She and three other writers, including her soon-to-be-husband, Percy Shelley, were staying at Lake Geneva in Switzerland during that summer of 1816, entertaining one another by telling ghost stories and competing to write the best one. Mary's vivid dream, in which she saw a "hideous phantasm of a man stretched out" and a scientist using a machine to try to bring him to life, inspired her story. She began to write it the next day.

something less exotic is going on. Recent research has shown that, more often than not, dreams embody organized and ordinary situations, while occasionally weaving in elements of the bizarre. Studies performed both in the lab and out of it suggest that these nightly pictures are generally less abstract.

"Dreams do enact—they dramatize. They are like plays of how we view the world and oneself in it," says G. William Domhoff, who teaches psychology and sociology at the University of California–Santa Cruz. "But they do not provide grandiose meanings." Domhoff bases his view on a study of themes and images that recur in a databank of more than 25,000 dreams that have been collected as oral

just as dull. They are overwhelmingly visual, and language is less important, and logic is less important."

I am a traveler carrying one light bag and looking for a place to spend the night. I...discover a hostel of a sort in a large indoor space big enough to house a gymnasium. I find a spot near a corner and prepare for bed. I think to myself, "Luckily, I have my high-tech pillow." I take out of my bag a light, flat panel about 8 by 10 inches and the thickness of a thick piece of cardboard. "It works by applying a voltage," I say. "There's a new kind of material which fluffs up when you apply a voltage." On the face of the panel is a liquid-crystal display with two "buttons," one labeled "on" and one labeled "off." I touch the "on" button with my index finger, and the flat panel magically inflates to the dimensions of a fluffy pillow. I lay it down on the ground and comfortably go to sleep.

Chuck, scientist
(from Dreambank.net)

If Chuck's experience is an example of logic gone to sleep, no wonder many dreamers wake up shouting, "Eureka!" Indeed, history is filled with examples of ideas that blossomed during sleep and eventually led to inventions, works of art, or simpler strokes of inspiration. Professional golfer Jack Nicklaus, for example, came charging out of a slump during the summer of 1964 after seeing himself in a dream using a different grip on his club, and adopting it. Exactly what happens to inspire such creativity is unclear, but the new technology is providing clues.

Crazy smart. Brain scans performed on people in REM sleep, for example, have shown that even as certain brain centers—the emotional seat of the brain and the part that processes all visual inputs—turn on, one area goes dormant: the systematic and clear-thinking prefrontal cortex, where caution and organization reside. "This can explain the bizarreness you see in dreams, the crazy kind of sense that your brain is ignoring the usual ways that you put things together," says Robert Stickgold, associate professor of psychiatry at Harvard and director of the Center for Sleep and Cognition at Beth Israel Deaconess Medical Center. "This is what you want in a state in which creativity is enhanced. Creativity is nothing more and nothing less than putting memories together in a way that they never have been before."

Putting memories together is also an essential part of learning; people integrate the memory of new information, be it how to tie shoelaces or conjugate French verbs, with existing knowledge. Does dreaming help people learn? No one knows, but some sort of boost seems to happen during sleep. Many studies by sleep researchers have shown that people taught a new task performed it better after a night of sleep.

A study of how quickly dreamers solve problems supports Stickgold's theory that the sleeping mind can be quite nimble and inventive. Participants were asked to solve scrambled word puzzles after being awakened during both the REM phase of sleep and the less active non-REM phase. Their performance

Paul McCartney and "Yesterday"

"I woke up with a lovely tune in my head," Paul McCartney recalled to his biographer, Barry Miles. "I thought, That's great. I wonder what that is?" He got up that morning in May 1965, went to the piano, and began playing the melody that would become one of the century's most popular and most covered songs: "Yesterday." At first, lacking lyrics, he improvised with "Scrambled eggs, oh, my baby, how I love your legs." While he really liked the tune, he admitted to having had some reservations: "Because I'd dreamed it I couldn't believe I'd written it."

improved by 32 percent when they worked on the puzzles coming out of REM sleep, which told researchers that the REM phase is more conducive to fluid reasoning. During non-REM sleep, it appears, our more cautious selves kick into gear.

Indeed, PET scans of people in a non-REM state show a decline in brain energy compared with REM sleep and increased activity in those dormant schoolmarmish lobes. Does this affect the content of dreams? Yes, say researchers from Harvard and the Boston University School of Medicine.

Since people should theoretically be less inhibited when the controlling prefrontal cortex is quiet, the team tracked participants for two weeks to see if their REM dreams were more socially aggressive than the ones they reported during non-REM sleep. The REM dreams, in fact, were much more likely to involve social interactions and tended to be more aggressive.

I had a horrible dream. Howard was in a coffin. I yelled and screamed at his mom that it was all her fault. I kicked myself that I hadn't waited to become a widow rather than a divorcee in order to get the insurance. I woke up feeling miserable, the dream was so icky.

Barb (from Dreambank.net)

To many experts, Barb's bad dream would be a good sign, an indication that she would recover from the sorrow of her divorce. A vivid dream life, in which troubled or anxious people experience

Problems find solutions; angst seems to be soothed; OUT-OF-THE-BOX IDEAS germinate and take root

tough emotions while asleep, is thought to act, in the words of neuroscientist Cartwright, as "a kind of internal therapist."

The enduring and vexing question is: How much do dreams say that is of value? Despite all the efforts to quantify, to measure, no one has an answer yet. But dreams have played a role in psychotherapy for over a century, since Freud theorized that they signal deep and hidden motivations. "A dream is the one domain in which many of a patient's defenses are sufficiently relaxed that themes emerge that ordinarily would not appear in waking life," says Glen Gabbard, professor of psychiatry and psychoanalysis at Baylor College of Medicine.

Sometimes, dreams can be a helpful diagnostic tool, a way of taking the emotional temperature of a patient. The dreams of clinically depressed people are notable for their utter lack of activity, for example. Eric Nofzinger, director of the Sleep Neuroimaging Research Program at the University of Pittsburgh medical school, has studied PET scans of depressed patients and has found that the difference in what goes on during their waking and sleeping states is far less dramatic than normal. On the one hand, he says, "we were shocked, surprised, and amazed at how much activity" there was in the emotional brain of healthy people during sleep. In depressed patients, by contrast, the vigilant prefrontal cortex, which normally is not very active during sleep, worked overtime. Never surrendering to the soothing power of dreams, the brain is physically constrained, and its dream life shows it.

Healing power. Is it possible that dreaming can actually heal? "We know that 60 to 70 percent of people who go through a depression will recover without treatment," says Cartwright, who has tested her theory that maybe they are working through their troubles while asleep. In a 2006 study whose results were published in the journal *Psychiatry Research*, she recruited 30 people going through a divorce and asked them to record their dreams over five months. Depressed patients whose dreams were rich with emotion—one woman reported seething while her ex-husband danced with his new girlfriend—eventually recovered without the need for drugs or extensive psychotherapy. But those whose dreams were bland and empty of feeling were not able to recover on their own.

Still, post-traumatic dreams don't replay the situations exactly, says Siegel. Rather, they usually rework certain aspects in a slightly distorted way.

I've sat straight up in bed many times, reliving it, reseeing it, rehearsing it. And it's in the most absurd ways that only a dream could depict...the one that comes to mind most, dreaming of a green pool in front of me. That was part of the radar scope. It was a pool of gel, and I reached into the radar scope to stop that flight. But in the dream, I didn't harm the plane. I just held it in my hand, and somehow that stopped everything.

Danielle O'Brien, air traffic controller for American Airlines Flight 77, which crashed into the Pentagon on Sept. 11, 2001 (in an interview with ABC News)

Many clinicians working with traumatized patients have found that their nightmares follow a common trajectory. First, the dreams re-create the horrors; later, as the person begins to recover, the stories involve better outcomes. One way to help victims of trauma move on is to encourage them to wake themselves up in the midst of a horrifying dream and consciously take control of the narrative, to take action, much as O'Brien appears to have done in her dream. This can break the cycle of nightmares by offering a sense of mastery. "If you can change

---

DID YOU KNOW...

The reason that coffee and high-test soda keep people awake at night is that caffeine blocks the action of adenosine, a neurotransmitter in the brain that controls sleepiness. That is, it prevents adenosine from doing its job: triggering the impulses that make people feel tired.

SOURCE: "BRAIN: THE INSIDE STORY," AMERICAN MUSEUM OF NATURAL HISTORY

MYSTERIES OF THE MIND

Delving Deeply Into Sleep

Scientists have struggled for decades to understand precisely why we sleep, an activity that takes up roughly a third of the human life span. There are about a dozen different theories, ranging from the body's need to repair wear and tear to the brain's work of establishing new neural connections, says Alon Avidan, associate director of the UCLA Sleep Disorders Center. Yet not one of these theories has been accepted so far as providing a full explanation.

Though it may not be well understood, the essential importance of sleep is underscored, Avidan says, by "almost a redundancy" of brain systems that ensure that the body gets its daily rest. For example, chemicals known as neurotransmitters send signals across the brain to prepare us for sleep. Two of these, the hormones serotonin and melatonin, work in concert to help regulate the body's wake-sleep cycle. During daylight hours, serotonin production increases, keeping us awake. As darkness falls, serotonin levels drop and melatonin production rises, stimulating a small area of the brain called the suprachiasmatic nucleus. The SCN signals that it's time for bed.

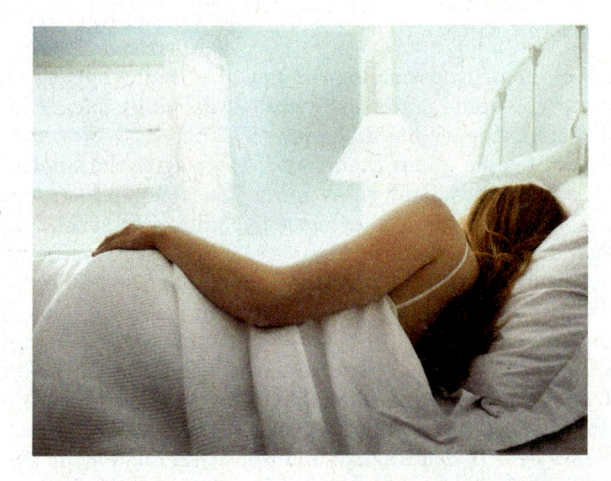

Disruption of these wake-sleep cycles can have serious health consequences. Laboratory rats have died of severe sleep deprivation, notes Clete Kushida, director of Stanford's Center for Human Sleep Research. In people, too little shut-eye has been linked to impaired immune function as well as higher rates of depression and obesity. Researchers have thus made it a priority to target the 80-plus disorders that wreak havoc on our slumber. One of the most dramatic is narcolepsy, believed to result in most cases from unusually low levels of hypocretins, two neuropeptide hormones that promote wakefulness. Narcoleptics can be overcome by sleep at any moment, putting them at risk of accident or even death, particularly if they are driving. Sufferers of sleep apnea, another common disorder, repeatedly stop and start breathing as the throat's soft tissues briefly close the airway to the lungs, the brain fails to signal the body to breathe, or both. Though the person awakes sufficiently to start breathing again, sleep quality is greatly diminished; this may increase the risk for heart disease and stroke.

Researchers have also focused on how sleep affects our ability to absorb new information. In a much-cited 2002 Harvard study, students learned to type a number sequence as quickly as possible. After a good night's sleep, they repeated the task and showed significant improvement in accuracy and speed. The brain seems to take "advantage of sleep to recognize and store what we did during the day," says Maria Bautista, a sleep disorder specialist at Georgetown University Hospital in Washington, D.C. Clearly, to be at your best, the old saw of getting an unbroken night of rest holds true—even if experts still search for the reasons why. *–Danielle Kurtzleben*

the dream content," says Harvard's Barrett, editor of *Trauma and Dreams*, "you see a reduction in all the other post-traumatic symptoms."

Cartwright recalls helping a rape victim who suffered from nightmares in which she felt an utter lack of control; together, they worked to edit the young woman's dreams of being in situations where she was powerless—of lying on the floor of an elevator without walls as it rose higher and higher over Lake Michigan, for example. "I told her, 'Remember, this is your construction. You made it up, and you can stop it,'" says Cartwright, who coached the woman to recognize the point at which the dream was becoming frightening and try to seize control. At the next session, the woman reported that, as the elevator rose, she decided to stand in her dream and figure out what was happening. The walls rose around her until she felt safe. For people like Cartwright's patient, creating a new ending in the waking mind can translate into the dreaming one.

A window? A royal road? A way for the brain to integrate today with yesterday? While definitive answers remain elusive, the experience of dreaming is clearly as universal as a heartbeat and as individual as a fingerprint—and rich with possibilities for both scientist and poet. •

By Marianne Szegedy-Maszak

MYSTERIES OF THE MIND

Divining the Secrets of the Soul

SCIENTISTS AND PEOPLE OF FAITH SEEK TO EXPLAIN OUT-OF-BODY EXPERIENCES

As head of the neuropsychology unit at University Hospital Zurich in Switzerland, Peter Brugger has interviewed dozens of people about their out-of-body experiences and read about many more, going back more than a century. "They cease to feel their bodies" and report "feeling like a gas that's evaporated from a bottle of ether," Brugger says. "Rather than sinking into the ground, you float up towards the ceiling."

For hundreds of years, such supernatural sensations have been a mystery—and to people of faith, a link to the divine. The sensations reported by people who have been to death's door and returned—the feeling that they were cut loose from their dying bodies to float upward, for example, or even reports that they watched as doctors struggled to revive them—seem like irrefutable evidence that there is something beyond our flesh and blood.

Ever since he first read about astral projection as a young man, Brugger has been interested in the phenomenon. Today, he believes that out-of-body experiences result from the brain's failure to combine the three senses that usually let us feel "in" our bodies: touch, vision, and our understanding of where our limbs are in space. In the last decade, Brugger and other researchers around the world have worked to uncover the science behind occurrences from near-death experiences to the transcendent feelings achieved through prayer or meditation. With the help of sophisticated brain scans, carefully placed electrodes, and old-fashioned psychological surveys, they are illuminating how the brain is responsible for a range of phenomena that seem mystical or mysterious to the average person.

DECOUPLING THE BRAIN from the body may help people function without panic in perilous moments

Studies by Swiss neurologist Olaf Blanke published in *Science* in 2007, for example, suggest that crossed wires in the brain may be to blame. Imagine the mind as a mosaic of tightly interconnected receivers, each responsible for processing different sensations. Multisensory processing regions combine the different streams of data into an overall sense of the body's physical place in the world. This process isn't perfect. Glitches often occur in daily life. Anyone who has ever suffered from car sickness has experienced this: The eyes, looking out the car window, tell the brain the body is moving, while the inner ear reports that the body is standing still. The conflict can have nauseating results.

Through decades of experiments and accidental discoveries—often when a specific brain injury results in unusual symptoms—scientists have been able to isolate the regions responsible for processing specific information. One of the most important is called the temporal-parietal junction, located where the lobes of the brain that handle auditory and spatial information meet.

In 2000, Blanke was treating a woman for epilepsy by administering electrical currents to her brain. When he stimulated a spot inside the temporal-parietal junction called the angular gyrus—a region known to mesh body awareness with visual information—the patient suddenly reported feeling like she was floating near the ceiling, looking down at her body. With a flick of a switch, Blanke could "return" her to her body.

In another series of carefully designed experiments, the neurologist had subjects wear virtual-reality goggles, which enabled them to see an image of themselves projected before them. When researchers touched the real person's back with a stick, the subject saw, through the goggles, the virtual double also being touched. Although the subjects did not feel themselves actually being projected into their virtual selves, they did report confusion about where they were located in space.

Brugger argues that the experiments demonstrated that out-of-body experiences are an illusion, a disruption in the multisensory processing region's ability to reconcile all the information it's receiving.

Others agree. "There is a coalescing of a lot of our senses—vision, motion, orientation, the position of our body. They all come together and get integrated" in the temporal-parietal junction, says neurologist Kevin Nelson of the University of Kentucky. "If that gets disrupted, we orient ourselves in our most dominant sense, our vision." That's why scientists think the final element of the out-of-body experience—the perception that one can see one's own body while floating above it—is an illusion created by the brain as it struggles to process the feeling of being separated from its physical self.

Survival mechanism. Though people may be reluctant to talk about them, these experiences are surprisingly common. In a study of 13,000 Europeans, nearly 6 percent reported having one. Yet under normal circumstances, feeling removed from one's self is an aberration. "The proper function is the feeling of being in our bodies. We cannot afford to be separated from our limbs, which explore our environment," Brugger notes.

Strikingly, however, there are ample reports of people in extreme situations experiencing just this. Having the sensation in a moment of life-threatening danger "is completely normal," Brugger says. High-altitude climbers who have found themselves in hazardous circumstances and people who have survived car wrecks all have reported seeing themselves from outside their bodies. Time seems to slow down for them, and the world becomes abnormally clearer, as though they are able to see more detail.

These sorts of experiences may not reflect brain malfunctions at all, but rather a sort of extreme adaptation. Researchers have speculated that decoupling the brain from the body could help people function without panic in perilous moments. "It could be a protective experience, to calm the person down" so he or she can better cope with the danger, Brugger says. "That's a biological advantage. It evolved to protect us."

In a 2007 study published in the journal *Neurology*, Nelson argued that the most extreme sort of out-of-body experience may be triggered for the same purpose. The lucky few who come to the edge of death and return often report a set of sensations so similar they seem almost clichéd: white light, a sense of floating out of the body, paralysis. Science is just beginning to probe the mysteries of near-death experience.

Neurologists have long known that the brain carefully monitors its blood supply. Any sharp change, from fainting to a heart attack, sets off a flurry of activity as the brain tries to jolt the body back to consciousness. "As the brain is losing blood supply, the natural response is high-amplitude electrical activity," Nelson says.

That phenomenon has made monitoring different brain centers at the moment of death difficult. "Physiological recordings are a great challenge," Nelson says. Thus, neurologists investigating near-death experiences have relied on interviews with survivors and guesswork to speculate on what is happening in the brain.

Nelson, author of the recent book *The Spiritual Doorway in the Brain: A Neurologist's Search for the God Experience*, thinks the answer may lie in the shadowy borderlands between dreaming and waking. In the study published in *Neurology*, he found that three fourths of people who reported near-death experiences also had sensations of leaving their bodies at some point. Most occurred during rapid eye movement sleep, also known as REM consciousness—the part of the sleep cycle when people dream.

Nelson says the part of the brain that's activated while dreaming also is activated when the brain senses a crisis like diminished blood flow. When animals or people lose significant amounts of blood and begin to go into shock, the brain suppresses the adrenaline system and may activate centers usually associated with dreaming. At the same time, it deactivates the processors by which people determine their place in the surrounding environment, effectively creating an out-of-body experience that may have a calming effect.

"The intriguing evidence is that REM consciousness and low blood flow explain almost the totality of near-death experience—light, tunnel, bliss, paralysis, the sensation of floating," Nelson says. "This is so deeply wired into the brains of humans that it must serve some very important, crucial adaptive purpose."

Changing the brain. As some researchers explore the involuntary responses that people may interpret as flights of the soul, others are gathering evidence for the brain's critical role in waking religious experiences. While the area that controls a person's spatial orientation is easy to pinpoint, the centers involved in religious experience are located all over the brain.

Working with Tibetan Buddhists, cloistered nuns, and charismatic Christians who speak in tongues, Andrew Newberg, director of research at the Myrna Brind Center of Integrative Medicine at Philadelphia's Thomas Jefferson University Hospital and Medical College, has looked at how the brain responds during prayer, meditation, and other religious experiences. He injects a radioactive dye into his subjects, then puts them in a single photon emission computed tomography (SPECT) scanner—a device usually used to take images of brain trauma, patients with Alzheimer's, or people suffering from seizures. The dye captures those parts of the brain that are active at the height of the religious experience.

Some of Newberg's findings were unsurprising, like the fact that the frontal lobes—responsible for sophisticated thought, a hallmark of the human brain—lit up. "If you are concentrating on anything, you turn on the frontal lobes," Newberg says.

But when people pray or meditate, other centers all over the brain that govern emotion and sensory inputs also begin to light up. Prayer and meditation draw on a more complex range of brain activity than run-of-the-mill thinking.

Interestingly, some differences between religious experiences also showed up. The nuns activated areas that had to do with speech, for example, while meditating Buddhist monks used the visual parts of the brain more. Most strikingly, one region stayed dark for both groups: the parietal lobe. This lobe "is involved with the sense of self," Newberg says, and is associated with sensory information and how people relate to the rest of the world. "In deep meditation practice where people lose their sense of self," he says, "we see a decrease in this kind of activity."

Through prayer, the people Newberg studied were able to free themselves, in a sense, from their bodies. What he is unable to answer is the larger question that brain imaging does not address. And that's the possibility, he acknowledges, that something else is occurring that science can't track. "It's very hard for us to know if the changes are created by the brain or if the brain is responding to something else out there," he says.

Newberg's research also has practical implications. His results suggest that religious practice could have physical benefits. The act of meditation, he says, literally changes the brain. Research has shown that longtime meditators have increased blood flow to important parts of the brain, and a study of older Americans who practiced meditation for two months showed that they had heightened activity in their frontal lobes and more efficient blood flow in other parts of the brain.

"People who regularly attend religious services" and follow other practices tied to their faiths "have better health outcomes," says Patrick McNamara, director of the Evolutionary Neurobehavior Laboratory at Boston University School of Medicine. They also "tend to resist the effects of aging longer." That makes belief a powerful system, he adds. And individuals who can recruit all those healing capacities are more likely to survive. ●

By Andrew Curry

> Through prayer, some people are able to **FREE THEMSELVES**, in a sense, from their bodies

MYSTERIES OF THE MIND

The Puzzle of the Psychopath

Brian Dugan was only 15 when he began his criminal career. The Illinois resident started with burglary, went on to arson and other offenses, and served a three-year prison sentence. After his release, his crimes escalated. In 1985 he pleaded guilty to the rape and murder of a 27-year-old woman and a 7-year-old girl, receiving life imprisonment. The same year, Dugan, who would be labeled an "extreme psychopath," told his attorney that he was also responsible for the 1983 murder of 10-year-old Jeanine Nicarico in Naperville, a Chicago suburb. Dugan promised to give a full accounting of the crime but only if he did not face the death penalty. The court accepted the guilty plea, but the prosecutor refused to bargain on the sentence.

When Dugan went before a Chicago court in the fall of 2009 for sentencing, his defense team called neuroscientist Kent Kiehl as a witness. Kiehl is an associate professor at the University of New Mexico and a principal investigator at the Albuquerque-based Mind Research Network, an independent nonprofit supporting the diagnosis and treatment of mental illness. He has also devoted years to using mobile functional magnetic resonance imaging machines, which track brain activity by monitoring changes in blood flow, to study the minds of psychopaths. Since 2006, Kiehl has taken fMRI devices into prisons and juvenile detention centers in New Mexico and Wisconsin, getting access to scores of research subjects.

Lack of engagement. Dugan's defense team had argued that, while there was no question of the killer's guilt, he should not be executed because he was unable to control his actions. Kiehl testified that Dugan's brain scan showed abnormalities: The amygdala, which governs emotions, and the orbitofrontal cortex, which controls rational decision-making and regulates impulsive behavior, did not show typical activity. In one test, when Dugan viewed unsettling and offensive images, he did not react normally. For example, a photo of a Ku Klux Klansman standing in front of a burning cross typically evokes horror and repugnance in most viewers, a reaction reflected by heightened activity in the amygdala. But Dugan's scan, like those of other psychopaths, did not show a significant level of activity, reflecting a lack of emotional engagement.

The description often used with psychopaths is that they "understand the words, but not the music." These individuals are not wired to feel empathy, guilt, or shame, and while they know something may be wrong, they just don't care. "You can sit a psychopath down and he'll say, 'I know it's not good if I kill people,'" says Kiehl. "But without anyone else there to act as an external brake, psychopaths just don't have that emotional switch that tells them stop." At the time of his arrest, Dugan told authorities, "I wish I knew why I did a lot of things, but I don't."

Despite Kiehl's testimony, the jury voted for the death penalty and Dugan is now awaiting execution at the Pontiac Correctional Center in Pontiac, Ill. Understanding that he was a psychopath apparently did not convince the jury that it mitigated Dugan's crime.

The psychopathic personality was first described in 1801, but it was American psychiatrist Hervey Cleckley who identified the traits now most associated with the pathology. In his 1941 book *The Mask of Sanity*, Cleckley noted that while psychopaths may often seem charming

Serial killer Ted Bundy showed evidence of psychopathy when he was a young child.

Brain scans of career criminal and murderer Brian Dugan did not show typical activity.

and intelligent, they are actually deceptive, manipulative, predatory, and completely lacking in empathy, emotion, or remorse.

Identifying them can be a challenge, however, since they are often so effective in assuming an ordinary facade. Today, many psychologists and neuroscientists use the revised Hare Psychopathy Checklist (PCL-R) developed by Robert Hare, professor emeritus of psychology at the University of British Columbia in Vancouver, and considered the gold standard. The checklist identifies 20 criteria, on which a person is scored from zero to 2. Most people score about 4 points on the continuum. Anyone who scores 30 or higher is considered a psychopath. Brian Dugan racked up 37 points.

Although serial killer Ted Bundy was never subjected to fMRI brain scans or the PCL-R checklist (he was executed before these tools became available), he is a prime example of a psychopath. Attractive, engaging, and intelligent, Bundy was a former law student who held a series of good jobs. (He even volunteered for a suicide prevention hotline.) But despite this veil of normalcy, while studying or employed, he kidnapped and murdered as many as 35 young women and girls from 1974 to 1978 in Washington, Idaho, Colorado, Utah, and Florida. Twice he managed to escape from county jails to continue his rampage before he was finally caught and sentenced to death for the murder of two college students and a 12-year-old girl. He was executed in 1989.

Bundy later became the motivation for Kent Kiehl's work, since the neuroscientist grew up in the killer's hometown of Tacoma, Wash., and would hear stories of his crimes. "I was fascinated by how somebody could live just down the street from me and end up like that," he says.

Psychopaths may "end up like that" because they start out that way. Bundy showed evidence of psychopathy when he was still a child. In her 1980 book, *The Stranger Beside Me*, author Ann Rule tells the story of Bundy's sleeping aunt waking up to find herself surrounded by knives from the family kitchen, with 3-year-old Ted standing by the bed, smiling eerily. The challenge for society is to identify psychopaths early, before they hurt anyone. Today, James Blair, chief of the affective cognitive neuroscience unit at the National Institute of Mental Health, is working with juvenile offenders, studying what he calls "callous-unemotional traits." ("Psychopath is a pretty heavy label to use with kids," he says.)

Like Kiehl, Blair uses an fMRI machine to scan his subjects' brains while they perform certain tasks. His results show that, even in children and adolescents, the amygdala and the orbitofrontal cortex process information differently. In response to a photo of a fearful face, the amygdala doesn't "light up" with activity, indicating a response to an

Snakes in the Executive Suite

Psychopaths make up about 1 percent of the general population. Not all of them are violent, but many people with psychopathic personalities are embedded in the workplace, where they often become a destructive presence, especially when occupying leadership positions. Paul Babiak, an organizational psychologist and coauthor with Robert Hare of *Snakes in Suits: When Psychopaths Go to Work*, notes that these executive psychopaths are generally charming, manipulative, and remorseless. Today's fast-paced, rapidly changing business environment—where embezzlement, insider trading, and mortgage fraud are everyday occurrences—makes it "easier for people with psychopathic tendencies to hide and thrive," says Babiak. The chaos creates opportunities for them to engage in manipulative schemes or abusive behavior.

While every bully is not a psychopath, most psychopathic executives are bullies, notes Gary Namie, director of the Workplace Bullying Institute in Bellingham, Wash. "The consequences of their behavior can be devastating, as talented employees are driven out of the company and relationships are damaged between a corporation and its clients and suppliers," he says.

To ferret out psychopathic executives, Babiak and Hare are refining what they call the Business-Scan questionnaire, designed for use by trained experts during interviews with executives being considered for leadership posts. The questionnaire helps identify traits like glibness, grandiosity, tendency to lie, manipulation, lack of empathy, and antisocial behavior. It is given to the individual and a supervisor.

While researchers still do not know why only some psychopaths become violent, the one trait they all share is that they care solely "for Number One," Babiak says. And for this, everyone else pays the price. – *C.G.*

MYSTERIES OF THE MIND

emotional cue. These young people either don't recognize or don't react to such cues like most people.

"In a nutshell, the amygdala is critical for specific types of emotional learning, such as whether a thing is good or bad," says Blair. "And on top of that, the orbitofrontal cortex helps you make decisions about what you've learned," he says. "Most of the individuals in jail are there because their decision-making system has been badly disrupted."

Rewards and regulation. Not only do psychopaths make poor decisions, they also appear to be wired to seek rewards at any cost. Joshua Buckholtz, a graduate student in neuroscience at Vanderbilt University in Nashville, is looking at how the brain's reward system might drive psychopaths. In research reported in March 2010, Buckholtz gave ordinary volunteers (not inmates or juvenile offenders) a test called the Psychopathic Personality Inventory (PPI), which is helpful in gauging the presence of important psychopathic traits among people not incarcerated. "The people who score highest on this measure of impulsive-antisociality are not your serial killers," says Buckholtz. "But they will have significant social and behavioral problems due to the presence of these traits. They are going to be what I sometimes call your Machiavellian mother-in-law, your bullying boss, or conniving co-worker." (Box, Page 46.)

The subjects were then given an amphetamine to stimulate the release of dopamine, a chemical important for motivating behavior to obtain potential rewards. Positron emission tomography (PET) scanning, which shows chemical activity in the brain, indicated how much dopamine had been released in response to the stimulant. "We found that the brains of those who scored the highest levels of psychopathic traits released almost four times as much dopamine as those who scored lower," he says. This exaggerated dopamine response may indicate that these individuals are focused on getting rewards, no matter what the consequences. Even if the "reward" is rape or murder.

In the second part of the experiment, the subjects' brains were scanned with fMRI while the participants performed a test in which they had an opportunity to win money. Those with stronger psychopathic traits showed much more brain activity while anticipating the monetary reward than other volunteers. "Heightened activity while preparing to obtain a reward may lead to a narrowing of attention on that goal and an insensitivity to potential threats or costs," says Buckholtz.

So does that mean psychopaths can't switch gears? Yes, says Joseph Newman, professor of psychology at the University of Wisconsin–Madison. In one of his experiments, participants were given a deck of cards and told that for every face card they turned up, they would earn money; for every non-face card, they would lose money. The deck was stacked so that lots of face cards showed at the beginning, then gradually diminished. Non-psychopaths gave up in disgust once they started losing, but psychopaths (identified by the Hare checklist) kept going, turning over almost all the cards in the deck (and losing all of the money that they had earned). "They were very upset when they were doing the task and losing money," says Newman. "Yet, they failed to stop and regulate their behavior. This study shows that there's some sort of information-processing deficit that impairs their regulation," he says.

This intense focus on a goal, coupled with a psychopath's lack of empathy, may explain the horrific murders recounted in Truman Capote's 1966 book

Provocative photos, such as a burning cross, elicit little reaction in the brains of psychopaths, reflecting a lack of social engagement.

In Cold Blood, which described how ex-cons Perry Smith and Richard Hickock broke into the Holcomb, Kan., house of Herbert Clutter in a planned burglary and murdered four family members in November 1959. Once the violence began, Smith and Hickock were not influenced by the victims' pleas. They were oblivious to "consequences for other people," says Newman.

Today, fictional monsters, like Norman Bates in *Psycho,* delight audiences by safely scaring them in the confines of a movie theater. Meanwhile, their real-life counterparts walk the streets, finding new victims to prey on. Experts are beginning to understand how psychopaths process emotions and information differently from other people. They will continue to study these predators in the hopes of finally finding ways to control them. But for now, psychopaths pose an ongoing, often unrecognized, threat to those around them. ●

By Cathie Gandel

The Brain Chemistry of Love

YOUR HEART IS RACING AND NOTHING ELSE MATTERS? BLAME THE STORM IN YOUR HEAD

Never mind that the song contends you can't hurry love—science says you don't actually need to. It takes all of a fifth of a second for that truly, madly, deeply feeling to register, says a report published in the October 2010 *Journal of Sexual Medicine*. That's faster than your average resting heartbeat.

Given the heart's popular role in romance, it may be a surprise that falling and being in love are actually more of a cerebral upheaval. Research into the wonders of passion has found that a concert of chemicals act on about a dozen parts of the brain all told (not to mention countless other body parts) to create that intense rush experienced as love. In fact, those neurotransmitters flood the system so fast that it appears "your brain knows before you do that you are in love," says Stephanie Ortigue, the Syracuse University assistant psychology professor who led the study of how fast the brains of people passionately in love lighted up when presented with the names or pictures of their significant others.

The physiological process all begins with a face. "Constituting only 5 percent of your bodily surface, it carries 95 percent of your allure," writes anthropologist David Givens in his book *Love Signals*. When the eyes fall upon an engaging countenance, sensory neurons fire up in the temporal lobes, located about ear level on either side of the brain. They send signals to the thalamus, bulb-shaped masses about 2½ inches long in the lower brain where sensations like sight and touch are processed. The thalamus then sends arousal and pleasure messages to the frontal lobe, the area of the brain that helps people decide on the best course of action. Testosterone and dopamine (the neurotransmitter that's involved in making you happy) work together to cause the lovesick

MYSTERIES OF THE MIND

person to become aroused. Palms get sweaty, the heart beats faster, a blush arises as these chemicals work on various parts of the body, including the sweat glands and genitals, and blood flow increases, explains Maryanne Fisher, a psychology professor at St. Mary's University in Halifax, Nova Scotia, and author of *The Chemistry of Love*.

A pea-sized clump of cells located at the base of the brain lights up, too, researchers revealed in a 2005 brain-scanning study of people newly in love. That cell clump, the ventral tegmental area, is part of the brain's reward and motivation systems, and, as the central refinery for dopamine, sends the powerful neurotransmitter "to higher regions, creating craving, motivation, goal-oriented behavior, and ecstasy," says Helen Fisher, an anthropologist at Rutgers University who conducted the MRI scans of people in early-stage romantic love. It's a bit harder to suss out "why Jack as opposed to Joe," she adds. "That's a metaphysical question." Meanwhile, surging norepinephrine and dopamine induce euphoria. Norepinephrine is a stimulant, which explains "why you might notice that the sky looks brighter, or music is louder, or basically that the world seems more alive," says Fisher.

The whole love cocktail, which also includes oxytocin, adrenaline, and vasopressin, essentially triggers several of the same systems that are activated when a person takes cocaine, says Arthur Aron, a social psychologist at Stony Brook University in New York, who has studied love. (Indeed, a 2000 study using functional magnetic resonance imaging at University College London showed that love activates the same areas of the brain as drug abuse.) Oxytocin, the cuddling chemical, is love's superglue (it bonds mothers to their infants), adrenaline is responsible for racing hearts and restlessness, and vasopressin raises blood pressure. Vasopressin also might encourage bonding tendencies in men. In male voles, at least, it created urges for bonding and nesting when injected and when naturally activated by sex.

One key structure operating during this initial thrill of love is the nucleus accumbens, a knot of neurons in the forebrain that is part of the body's reward system. Sex, a good dose of chocolate, and a hit of cocaine are all the sort of pleasurable trigger linked with an increase in dopamine in this region (and it's been studied for its role in addictions). Next, the love signals jolt the two shrimp-sized caudate nuclei near the center of the brain that house 80 percent of all brain receptors for dopamine. When the caudate is flooded with dopamine, it sends signals calling out for more. "The more dopamine you get, the more of a high you feel," says Lucy Brown, a professor in the department of neurology and neuroscience at Albert Einstein College of Medicine in New York. As the reward system of the brain kicks into high gear, "anything that the object of one's lust does, like their touch, or even a thought of them," causes that intense reaction, says Fisher. Signs that this is happening include a focused attention, tons of energy, mania, elation, and "needing, craving more," she says. At the same time, the dopamine deluge lowers serotonin levels by up to 40 percent. It's serotonin that's in the driver's seat when it comes to regulating impulses, unruly passions, and obsessive behavior, and a reduction can enhance the feeling of being out of control. Lowered serotonin levels, which also occur in people with obsessive-com-

Sex, a good dose of chocolate, a hit of cocaine—all are linked with INCREASED DOPAMINE in the nucleus accumbens

pulsive disorders, "may explain why we concentrate on little other than our partner during the early stages of a relationship," says Domeena Renshaw, professor emeritus of psychiatry at Loyola University Chicago Stritch School of Medicine.

While all this neuronal firing is going on, the amygdala and the anterior cingulate cortex—the worry and caution centers—practically go into hibernation. So negative emotion is dulled, as is critical judgment. Says Brown: "When you're in a relationship, you're aware of the other person's flaws, but your brain is telling you it's OK to ignore them." In other words, science is now revealing that centuries of poets have had it right: Love is blind. ●

By Courtney Rubin

DID YOU KNOW...

Listening to certain types of music releases pleasure-inducing dopamine in the brains of music lovers. Research has shown that one reason people particularly enjoy favorite tunes is that the songs have become linked to their fond memories, especially from the teenage years.

SOURCES: MCGILL UNIVERSITY STUDY, *NATURE NEUROSCIENCE*, JAN. 9, 2011; *THIS IS YOUR BRAIN ON MUSIC* BY DANIEL LEVITIN

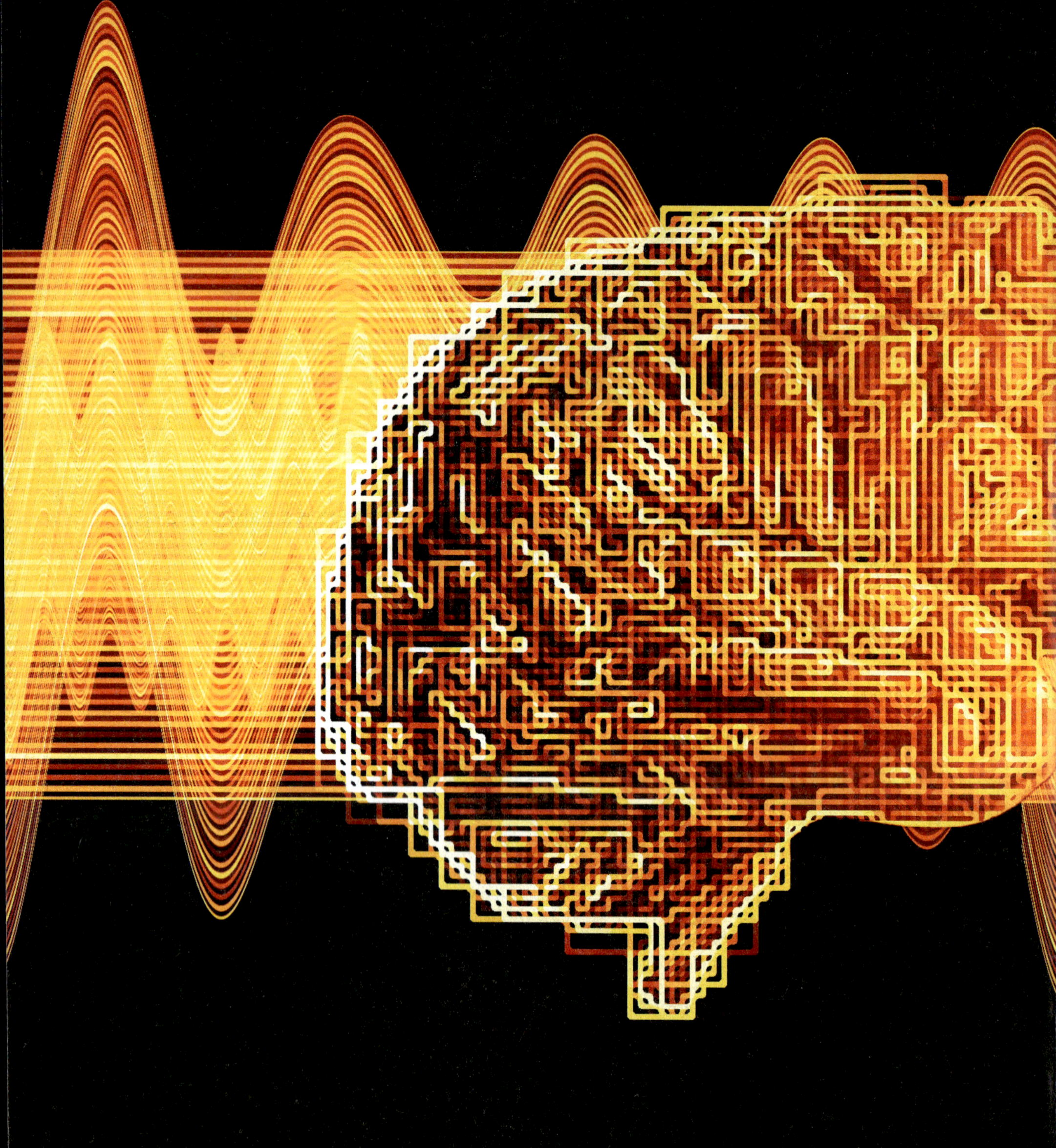

3. Using Your Brainpower

The latest on keeping your brain sharp **52**

Steps you can take to build a better memory **57**

The ADHD advantage **60**

Making the most of dyslexia **63**

Getting a handle on pain, it turns out, means understanding the brain's central role **66**

How to unleash your creative genius **68**

USING YOUR BRAINPOWER

Keeping Your Mind Sharp

According to Nintendo, Marian Conte's brain weighs 1,100 grams. "That's up from 800 grams when I started playing," jokes Conte, 55, a real estate agent from Hamilton, N.J., who added the video game Big Brain Academy to her fitness regimen a few years ago. The better she scores on brain teasers, the larger her fictional brain. Since Conte's mother died of complications from Alzheimer's disease in 2003, she's been trying to guard herself against the disorder any way she can, embracing crossword puzzles, fruits and vegetables, and a new genre of high-tech workouts that aim to slow cognitive loss. This particular game makes no such claim. But regular play certainly can't hurt, Conte figures. "I want to do any little thing I can to protect my brain."

If a Nintendo score isn't solid evidence, science has increasingly suggested that certain efforts may pay off—though a comprehensive review last year of the literature on ways to protect cognitive function, sponsored by the National Institutes of Health, found many studies seriously wanting. Just within the past few years, several groups of researchers have put forth provocative findings that it's possible to slow the age-related declines in memory, mental speed, and decision-making that affect most people. A team from the Mayo Clinic and the University of Southern California revealed that one mental training program appeared to improve older people's cognitive performance by as much as 10 years. Harvard researchers found that long-term use of beta carotene supplements delayed cognitive decline by up to a year and a half. And there's reason to believe that "exercise is the single best thing you can do for your brain," says John Ratey, an associate clinical professor of psychiatry at Harvard Medical School and author of *Spark: The Revolutionary New Science of Exercise and the Brain* (box, Page 54).

"Some of the myths about the brain—that it was not changeable, that there was nothing you could do about cognitive decline—have really been dispelled in the past 10 years," says Lynda Anderson, director of the Healthy Aging Program at the federal Centers for Disease Control and Prevention, whose bold goal is "to maintain or improve the cognitive performance of all adults." The potential payoff is enormous. Alzheimer's now afflicts more than 5 million people in the United States—more than double the number in 1980—and could reach as many as 16 million by 2050. "Statistics show if we could delay the onset of Alzheimer's by five years, the number of people with the disease would be cut in half," says Yaakov Stern, a cognitive neuroscientist at Columbia University.

What are you up against? The physical changes that mark normal aging start in early adulthood but become especially marked after age 60 or so. The brain shrinks, losing around 0.5 percent to 1 percent of its volume each year after that age threshold; brains with Alzheimer's shrink about twice as fast. The effects are greatest in the prefrontal cortex, the seat of executive function (which includes working

Alzheimer's disease now afflicts more than 5 MILLION AMERICANS and could reach 16 million by 2050

USING YOUR BRAINPOWER

me more intuitive: When someone sends me a written proposal, rather than dwelling on detailed facts and figures, I find that my imagination grasps and expands on what I read."

Rewiring the brain for words. You may be thinking at this point, "All these abilities sound great, but you still need to be able to read well to make your way in society." This is very true, and there is encouraging news from neuroscience on this subject. In a series of studies completed over the past five years, researchers have shown that the areas of the brain that good readers use can be activated in dyslexics by developing their phonological skills.

For example, researchers from Georgetown University Medical Center in Washington, D.C., collaborated with colleagues at Wake Forest University School of Medicine in Winston-Salem, N.C., on a study comparing two groups of dyslexic adults. One group completed an eight-week intervention using a multisensory, phonologically based remediation program administered by Lindamood-Bell Learning Processes; the other group received no intervention.

Results showed that the group undergoing the intervention read more proficiently and showed changes in their brain scans, indicating that their patterns of activation had become more like those of normally reading adults. "People in this study showed us that it may never be too late for adults who want to improve their reading skills," said Lynn Flowers from Wake Forest, the senior author who had followed this group of dyslexics since the 1980s. There are a number of reading remediation programs for adult readers recommended by Shaywitz, including the Wilson Reading System and Lexia Reading.

Making the most of a dyslexic brain. Though dyslexics frequently enter professions where visual-spatial or entrepreneurial abilities are required, others flourish in jobs relying heavily on the written word. Dyslexic authors, past and present, include novelist John Irving, screenwriter and producer Stephen Cannell, and mystery writer Agatha Christie. The prominent trial lawyer David Boies is dyslexic. So was Woodrow Wilson, who was the president of Princeton University before becoming president of the United States.

Many eminent psychiatrists and physicians are also dyslexic, including Cleveland Clinic CEO Delos "Toby" Cosgrove. In fact, people with dyslexia can succeed in any career if they have the drive and really put their minds to it, whether they want to be play-wrights or surgeons, engineers or entrepreneurs.

Often, dyslexics who settle on a suitable career find it useful to put together a human resource network to help them get things done, especially in their areas of deficiency. John Chambers, CEO of Cisco Systems, has his staff prepare three-page summaries of his reading material, with major points highlighted in yellow. He relies on his wife to help him navigate a phone book.

A boon to dyslexics (as well as to the blind) is

> **People with dyslexia CAN SUCCEED IN ANY CAREER if they have the drive and put their minds to it**

text-to-speech software, which uses a device to scan printed text and then translates the material into a digital "voice" that "speaks" to the user. In addition, there are reading software programs that use text-to-speech technology. Users read text on the screen and can get help with individual words or blocks of text by clicking on the highlighted material and having the computer read the material back. E-book readers like Kindle and Nook offer a wide variety of audiobooks.

If dyslexics are riding the wave of the future, then we need to alter our educational systems. Visual-spatial strategies should be employed so that children with dyslexia can harness their other abilities and strengths. Accommodations should also be made in the workplace, so that dyslexics can use the exciting technologies now available to adapt their work to their needs and styles.

Perhaps the day will soon come when the dyslexic is no longer seen as disabled but is looked upon more as a different kind of information processor whose out-of-the-box brain is a decided asset to the world. ●

Excerpted from *Neurodiversity: Discovering the Extraordinary Gifts of Autism, ADHD, Dyslexia, and Other Brain Differences* by Thomas Armstrong (Da Capo Press, 2010).

DID YOU KNOW...

Learning more than one language can have important cognitive as well as cultural benefits. Bilingual adults have been shown to have thicker gray matter than their monolingual peers. Other studies suggest that speaking multiple languages might help improve focus and delay dementia.

SOURCES: SOCIETY FOR NEUROSCIENCE; *THE HUMAN BRAIN BOOK* BY RITA CARTER

USING YOUR BRAINPOWER

The Epicenter of Pain

Neuroscientists at Stanford University School of Medicine wanted to know whether the power of romance could overcome physical pain. They recruited 15 students who were early in an intense relationship and subjected their left hands to thermal pain—enough to hurt, but not to harm them. When the students focused on a photograph of the person they loved, they felt less of a burning sensation than when gazing at a picture of an equally attractive acquaintance. In their peer-reviewed, 2010 study, the researchers concluded that romantic feelings activated regions of the brain involved with endogenous opioids, the body's natural pain relievers, and dopamine, a neurotransmitter tied to cravings and rewards.

The findings don't suggest people should seek out love affairs, notes Sean Mackey, chief of Stanford's pain management division. But they do indicate that engaging in "emotionally salient and rewarding experiences" can trigger the brain to provide relief, he says.

Since the brain is the epicenter of pain in the body, natural and medicinal strategies to relieve discomfort often focus there. For example, when you stub your toe, skin receptors send electrical signals through nerve fibers to the spinal cord and then up to the brain. Some of these fibers run like insulated telephone wires and carry the signals rapidly; others connect through web-like neural connections and travel more slowly.

In the case of a stubbed toe, the signals move rapidly up insulated nerve fibers to the brain's thalamus, which acts as a relay station and directs them to the sensory cortex. The signals are then interpreted by the brain as a sharp pain. The slower impulses, traveling through the web-like neural fibers, become a throbbing ache felt through the entire toe, a warning to treat the area gingerly while it heals.

As all this is happening, the brain also sends messages to the spinal cord that can amplify or dampen the pain. For example, a football player might jam his knee, but barely notice amid the excitement of a game. Most people can recall similar experiences, such as discovering bruises that they can't remember getting. These phenomena occur because of the brain's ability to filter out pain while the body engages in more urgent matters. Known as descending neural inhibitory control, it varies from person to person, says Andrea Trescot, a pain specialist in Jacksonville, Fla.

The brain's emotional control center, the limbic system, adds a second dimension, responding to how a person feels about pain or interprets its significance. David Kloth of Danbury, Conn., a pain specialist and spokesman for the American Society of Interventional Pain Physicians, notes that genetics, upbringing, and the cultural practices of various ethnic groups can all affect how pain is felt.

For example, Kloth says, people who were coddled as children by their parents every time they had small injuries may react strongly as adults to the slightest discomfort. This conditioning may contribute to the development of psychosocial problems such as anxiety, depression, or stress, which can amplify pain.

Temporary or acute pain, caused by minor ailments like a sprain or a burn, usually resolves itself after the affected region heals. When an acute situation goes unresolved or causes a malfunction in the nervous system, however, the pain cycle becomes self-perpetuating. In these cases, diagnosis and treatment can be challenging because the pain signals may reverberate throughout the nervous system, disguising the original source.

Mackey says research has shown that chronic pain should be viewed similarly to chronic diseases, like diabetes, since it likely will require the same kind of sustained, comprehensive treatment plan. But understanding and utilizing some of the options for activating the brain-nerve relationship remains central to the process:

The medicinal approach. While the goal should always be to find the root cause of pain, medication can be used to provide short-term relief. Different drugs address various aspects of the neural

> The brain has the power to **FILTER OUT PAIN** while the body engages in more urgent matters

pathways and are effective in blocking pain signals from getting to the brain, says Charles Inturrisi, professor of pharmacology at Weill Cornell Medical College in New York City. These are the major categories:

- Anti-inflammatory medications, ranging from aspirin to nonsteroidal drugs and steroids, act on the nerves that detect pain on the periphery of the body. They generally succeed best with problems like osteoarthritis, where bones wear on each other, or an inflamed wound.
- Anti-seizure drugs may correct the spontaneous misfiring of sensory neurons, which can occur with herniated disks, headaches, or chronic regional pain syndrome, also known as CRPS.
- Antidepressants build up the brain's ability to block descending pain signals and work for many conditions involving nerve injury. (They do not require the person to be emotionally depressed.)
- Opioids and some synthetic narcotics tone down ascending sensations of pain and amplify descending inhibitory signals. Doctors use opioids with care. They can be addictive, since they activate the brain's ventral tegmental area, which is related to the limbic system and rewards people during pleasurable activities like eating or having sex. Opioids are most appropriate for acute pain, but may be used carefully with chronic pain, Trescot says.

A homeostatic process. A form of medical acupuncture being practiced increasingly in the United States blends Chinese approaches, developed over 1,500 years, with a Western understanding of neurophysiology, says Gary Kaplan of McLean, Va., who is board-certified in medical acupuncture, family medicine, and pain medicine. Acupuncturists place fiber-thin needles at varying depths, often between ¼ and 2 inches, at carefully mapped points on the body, depending on the diagnosis. The needles arouse the nerves and release endorphins, which activate opioid receptors in the spinal column and brain, relieving the pain. Acupuncture stimulates the release of neurotransmitters, and the increase in their production seems to be progressive with long-term treatment. "It's a homeostatic process," says Kaplan. The goal is to bring the body back to its normal ability to self-heal.

Mind over matter. Research has shown that people can activate their descending pathways to block pain by distracting themselves with work, family, or hobbies they enjoy. Studies also have found that clinical hypnosis can help patients learn how to alter activity in specific areas of the brain, says Mark Jensen, a clinical psychologist and professor at the University of Washington Department of Rehabilitation Medicine. For example, hypnotic suggestions can reduce patient discomfort by decreasing activity in the anterior cingulate cortex, the area of the brain that processes the emotional response to pain. Such suggestions can also target the sensory cortex,

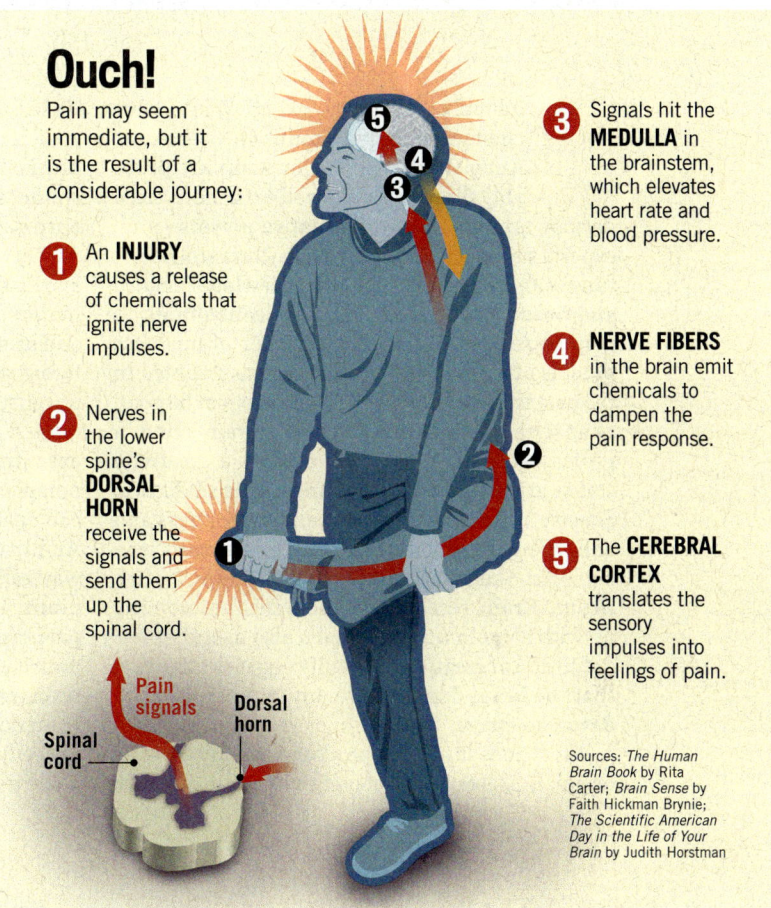

Ouch!

Pain may seem immediate, but it is the result of a considerable journey:

1 An **INJURY** causes a release of chemicals that ignite nerve impulses.

2 Nerves in the lower spine's **DORSAL HORN** receive the signals and send them up the spinal cord.

3 Signals hit the **MEDULLA** in the brainstem, which elevates heart rate and blood pressure.

4 **NERVE FIBERS** in the brain emit chemicals to dampen the pain response.

5 The **CEREBRAL CORTEX** translates the sensory impulses into feelings of pain.

Sources: *The Human Brain Book* by Rita Carter; *Brain Sense* by Faith Hickman Brynie; *The Scientific American Day in the Life of Your Brain* by Judith Horstman

which determines components of pain such as its intensity and what it feels like. A variety of disciplines are examining the use of hypnosis, which has been employed in dental procedures and studied as an aid to reduce pain in women with breast cancer.

Stanford's Mackey is currently exploring how to teach patients to manage their own pain. In 2006 he used functional magnetic resonance imaging scans to successfully train patients to consciously increase or decrease activity in the regions of their brains that processed the pain they experienced. While the research continues, Mackey hopes the approach will eventually be refined sufficiently to allow for its widespread use. ●

By Marshall Allen

USING YOUR BRAINPOWER

How to Unleash Your Genius

IT TURNS OUT THAT THE CREATIVE PROCESS INVOLVES BOTH ART AND SCIENCE

Suddenly, creativity is big. While your chances of making millions as the next Andy Warhol or Taylor Swift are probably slim, you could well earn more these days by tapping into your creative powers—and, experts say, you'll be happier, too. Numerous Fortune 500 companies, including Hewlett-Packard and Sears, have hired creativity consultants to help boost innovation. The number of business schools offering creativity classes has doubled in the past five years. "It's not enough to just be good at analytical evaluation," argues Yoram Wind, a professor of marketing who teaches a creativity course at the University of Pennsylvania's Wharton School. And creative activity can relieve stress and enhance your mood, according to Harvard psychologist Shelley Carson, author of *Your Creative Brain*. Brain researchers theorize that coming up with something novel that's also useful—their definition of creativity—so fully engages attention that the brain doesn't have any resources left to devote to stress.

What does it take to produce something truly original? The notion that creativity is the province of right-brain, left-handed artsy types is outdated, says Daniel Goleman, a psychologist and author of *Emotional Intelligence* (story, Page 32). "The creative brain state accesses a whole range of connections throughout the brain," he says. In fact, the latest research suggests that less than a second before the proverbial light bulb switches on, a spike in gamma brain waves appears to bind cells in several regions of the brain into a new neural network.

But fresh insights don't usually just spring forth. Whether you're mulling the next iPad or a solution to world hunger or just an artful way to rearrange your living room furniture, the creative process "is less about talent and more of a broad-based style of thinking that we all can learn," says Carson. The key is to approach it as a step-by-step process similar to proving a mathematical theorem. Leave out a step, and that stroke of genius may be elusive.

STEP 1: Absorb. Before you can come up with a brilliant idea, you need to openly receive information from the world around you, Carson says, and examine what's happening in your field of interest without judging it. Consider that the best novelists are the most avid readers, and that IBM invited computer hackers to speak to company executives about software innovations. "A fresh perspective can be very powerful," says Wind; it can enable you to examine all sides of a problem. And listening in a receptive and nonjudgmental way generates low-frequency alpha waves in the brain, allowing information stored in areas that perceive and freely associate to rise into conscious awareness and inspire a creative insight. You can train yourself to be more open to new ideas, Carson contends, by paying attention to what's happening in the moment, a practice called mindfulness. Set aside five minutes to simply experience the world around you: the colors, the sounds, the temperature, the sun's reflection, the approaching darkness.

STEP 2: Envision. Tapping into rich mental imagery, a practice that kids and daydreamers excel at, also inspired Einstein, who determined that the speed of light was constant by visualizing a light beam racing down a railway track and passing, at the same speed, a woman on a moving train and a man standing still on the platform. Carson recommends giving your visualization skills a workout for five minutes a day. Close your eyes and imagine you're taking a video tour of your bedroom. Enter the room and turn to the left, seeing the wall adjacent to your doorway in your mind's eye. Examine any furniture, windows, or drapes against this wall. Do the same for the three other walls. Next, look at the bed. Is it made? Are there clothes strewn on the floor? Then take a tour of your closet.

Japanese imaging researchers have found that

> You could well earn more by tapping your **CREATIVE POWERS**—and, experts say, you'll be happier, too

alistic and practical terms about what *will* work instead of how you'd like it to work. "It's the perfect place to take a fanciful idea, to flesh it out and make it practical," Carson writes in her book. People who doubt their creativity often get tripped up here by jumping ahead to "evaluation" (see Step 5), allowing self-doubts to shoot their ideas down. Carson recommends giving yourself verbal commands like "don't go there" to stop such thoughts, or using the visual image of a stop sign to push the thoughts away.

STEP 5: Evaluate. It's at this point in the creative process that a thoughtful and critical judging of your idea becomes necessary; "the evaluate brainset is where you want to be when you're deciding which idea or solution to implement," says Carson. She admits to being "a horrible evaluator" herself, however. "I'm an absorber," she says. "I come up with a lot of ideas, then they just sit there or dissolve from my memory the way my dreams dissolve." (She now carries a notebook and pen or a digital voice recorder to record her ideas for later pondering.) Practice using your evaluate mindset by making lists of your 10 favorite books, movies, restaurants, acquaintances, or memories, say, and ranking them in order of your preference.

STEP 6: Dive in. After you've figured out how to implement your idea, completely immerse yourself in arriving at the goal. Ideally, says Carson, you'll enter a brain activation state that Claremont Graduate University psychologist Mihaly Csikszentmihalyi has described as "flow," in which you lose all sense of time and self as you engage fully and spontaneously in responding to the challenge. To train your brain to get more easily into this mindset, spend time doing activities you really enjoy, and think of ways to make other tasks more fun and challenging. Can you load the dishwasher in under two minutes? Can you write that E-mail in less than 30 seconds? Or, as you dust your furniture, purposefully think back with pleasure to how you acquired each piece. Lastly, it often helps to raise your standards. Take pride in the oil change you're performing on your car—and do a little detailing while you're at it. •

By Deborah Kotz

when the "envision brainset" is properly activated, a network connecting the reasoning center in the brain's right hemisphere to the center in the left hemisphere dedicated to processing information from the senses has a burst of heightened activity. Visualizing often works best after an intense bout of physical activity, a time when your brain is ready for a snooze and when daydreams normally occur, Carson says.

STEP 3: Connect. After fully researching all the possibilities, encourage connections to happen by thinking about something else. "Distract yourself by taking a walk or reading a book," advises Goleman. "Trying to force an insight can stifle it." Mozart claimed he came up with his ideas for symphonies while taking carriage rides after a long repast. Getting outdoors into nature is a great way to distract the mind, suggests Carson. "Defocusing" lowers activity in the prefrontal cortex, where decisions are made and dangerous risks avoided, while heightening activity in the right temporal lobe. "This area of the brain understands the language of the unconscious, the logic of dreams, myths, art," says Goleman. "It helps put your ideas together in a novel organization."

STEP 4: Reason. Now you're ready to enter what Carson calls the "reason mindset," and think in re-

MORE @ USNEWS.COM Find out how to fan your child's creative spark at **www.usnews.com/creativekids**.

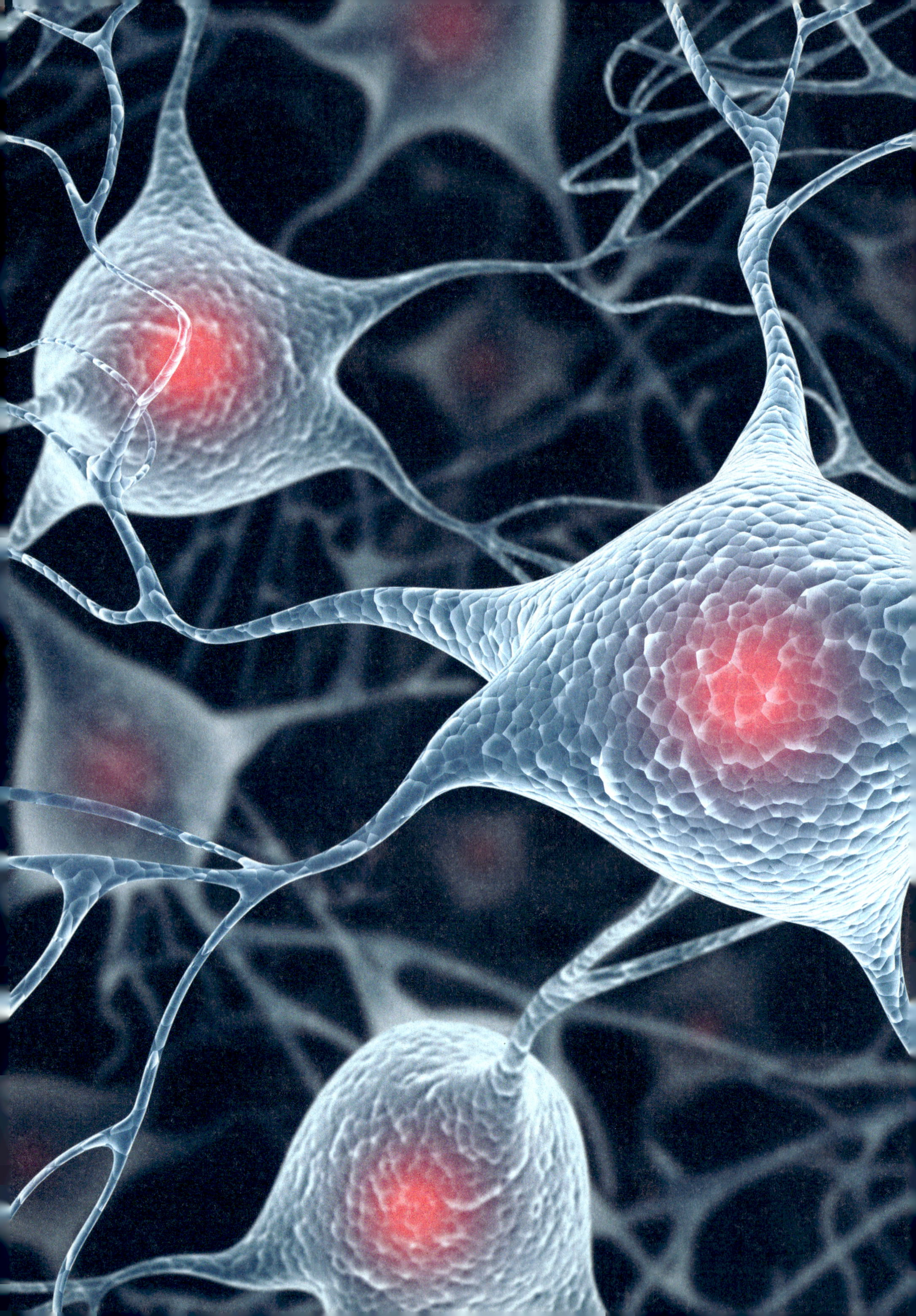

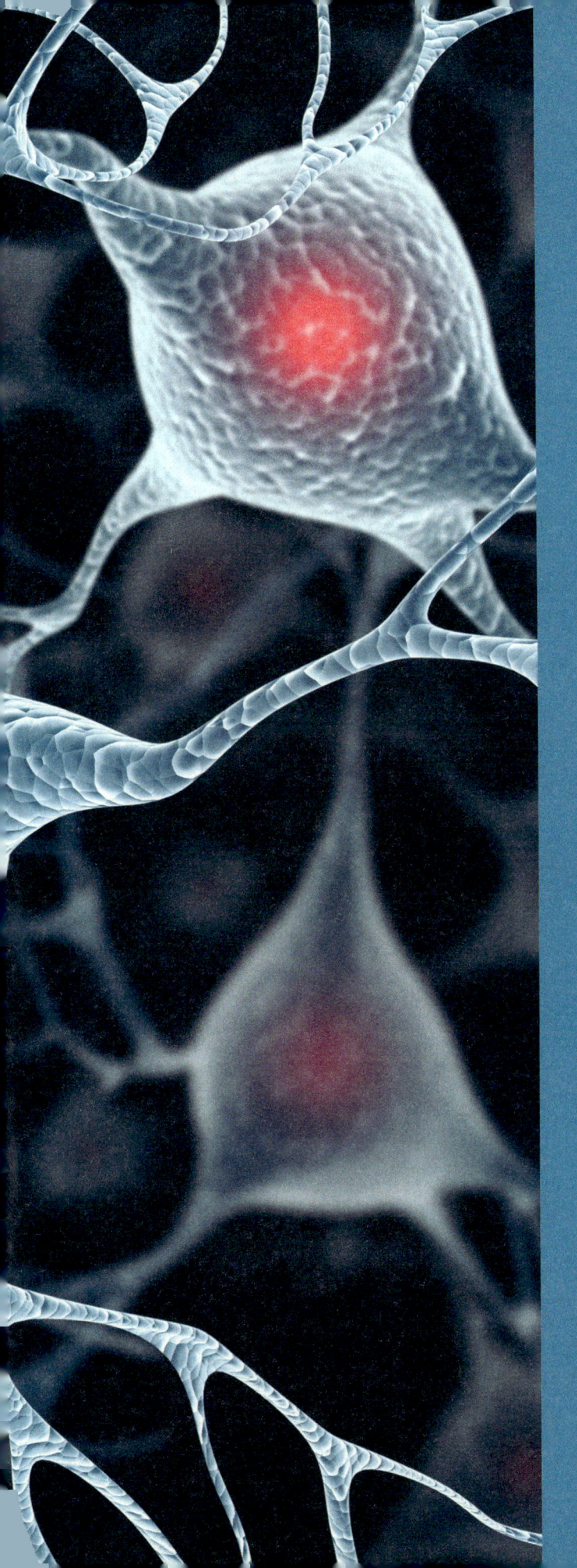

4.

Healing the Brain

Where does medical science stand? The latest on Alzheimer's disease, autism, stroke, brain cancer, and more **72**

Electroconvulsive therapy and other non-drug defenses against depression **79**

Exclusive: Inside the operating room with the pioneering neurosurgeons of the University of California, San Francisco Medical Center **82**

Exploring Medicine's Frontiers

THE LATEST DISCOVERIES, FROM ALZHEIMER'S TO AUTISM TO STROKE

The brains of the mice that Bradley Hyman keeps in his sprawling lab at an old naval base in Boston offer a window, literally and figuratively, into the mysterious damage that causes Alzheimer's disease. When each mouse reaches a few months of age, one of the lab workers carefully creates an opening in its skull and places a tiny glass window over the hole. Day after day, week after week, a powerful microscope is trained on the brain, searching for ugly clumps of sticky protein fragments like those that litter the brains of elderly people who have died of Alzheimer's. "It's like time-lapse photography," says Hyman, director of the Massachusetts Alzheimer's Disease Research Center at Harvard Medical School. When the ugly plaques appear—and they always do, as the mice carry genes engineered to produce them—nearby brain cells begin to wither and die, interrupting the flow of information. Next, waves of cells die off.

Hyman's microscope is one of several new technologies that promise to revolutionize the struggle to understand and beat Alzheimer's, which now afflicts more than 5 million Americans. Worldwide, a staggering 1 percent of all economic output is spent caring for and treating people with it and other types of dementia, according to Alzheimer's Disease International, the umbrella group of Alzheimer's associations around the globe. Meanwhile, just four drugs have been approved by the Food and Drug Administration to battle the disease, and all address symptoms only, not the poorly understood causes. Over the past decade, billions of dollars have been poured into researching drug after initially-promising drug, and nearly all have been disappointing in large clinical trials.

"When you can watch the brain over time, we see now, we didn't have the details right," says Hyman, who holds out great hope that his microscope studies will help correct that. One assumption has been that the plaques themselves, accretions of a protein fragment called beta amyloid, harm the brain. Instead, it appears that the individual sticky strands that eventually form the plaques damage neurons, and that the plaques are a sign of a brain long under siege.

For years, researchers have debated whether the brains of people who develop Alzheimer's produce excess beta amyloid or they're simply bad at clearing it. In December, researchers at Washington University School of Medicine in St. Louis provided strong evidence for the latter theory. They measured radioactively labeled beta amyloid visible in the spinal fluid of healthy older adults and people with Alzheimer's. Both groups appeared to produce the same amount, but the ill individuals cleared the substance from their brains into their spinal fluid

> Billions have been poured into researching **ALZHEIMER'S DRUGS,** and nearly all have been disappointing

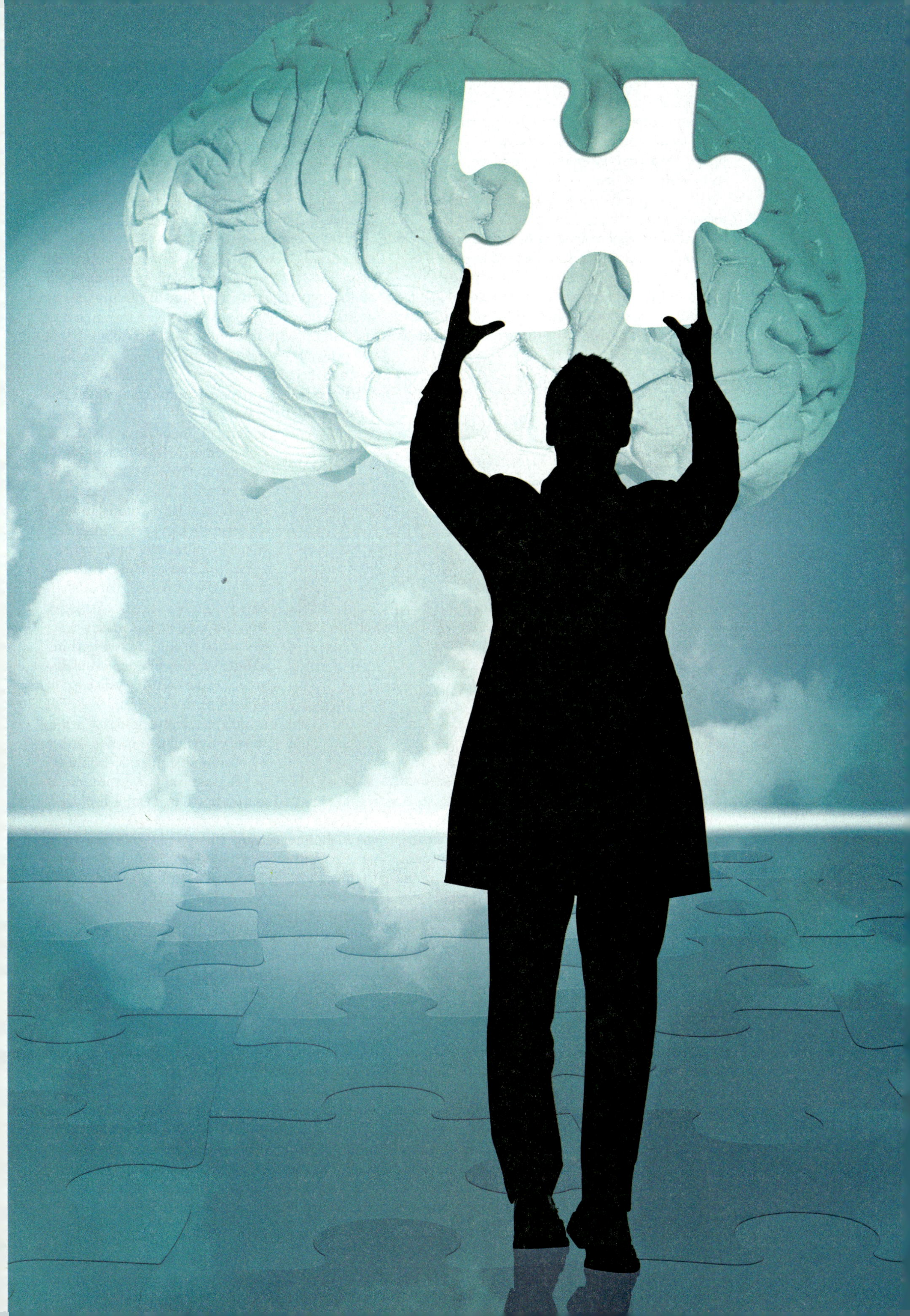

at a rate about 30 percent slower. Moreover, since autopsy studies find that some people with no cognitive symptoms of Alzheimer's carry a substantial plaque load, the body may possess a varying capacity to withstand beta amyloid's assault.

Following on the success of periodic cholesterol testing, which has revolutionized heart care, researchers are experimenting with measuring beta amyloid via brain scans, spinal taps, and blood tests. Early results offer hope that, someday, physicians will be able to screen the middle-aged for the hallmarks of pre-Alzheimer's. "It's critical to identify people at risk," says Reisa Sperling, who is using brain PET scans of beta amyloid at Brigham and Women's Hospital in Boston to study the impact in people who are not showing symptoms. She's concerned that right now drug treatment starts five or 10 years too late. An FDA advisory committee recommended in January that the agency approve a PET scan that could be helpful in diagnosing people who already have the plaques.

Of course, without effective drugs, early detection offers little solace. Most of the recently failed drugs aimed to interrupt production of beta amyloid by blocking the enzyme that produces it. However, researchers have opened a new front in the battle, targeting the synapses between neurons, which new research shows may be the first structures to deteriorate. Another approach, based on the theory that the body must have natural defenses if most people don't get Alzheimer's, seeks to train the immune system to attack beta amyloid. But all of the new drugs have been given to patients who already show cognitive symptoms. Adrian Ivinson, director of the Harvard NeuroDiscovery Center, which focuses on degenerative brain diseases, thinks some of the failed drugs should be retested in asymptomatic individuals who show beta amyloid on brain scans or in spinal fluid tests. "The implication of all this work is we have to get [the drugs] into people before they're patients," he says. "They have a silent disease."

What about prevention? There is accumulating evidence that exercise, eating fish or other sources of omega-3 fatty acids, and remaining intellectually and socially engaged throughout life reduce the risk of Alzheimer's. But last year a panel assembled by the National Institutes of Health concluded that the evidence on lifestyle interventions is inconsistent and inconclusive, pretty much across the board. "There are suggestions that some things might be effective, but there isn't strong, high-grade evidence for any of them," says Neil Buckholtz, chief of the dementias of aging branch of the National Institute on Aging. The NIA is now running several large studies to see if exercise, diet, or social or intellectual engagement will reduce risk.

> About **1 PERCENT** of global economic output is spent dealing with Alzheimer's and other dementias

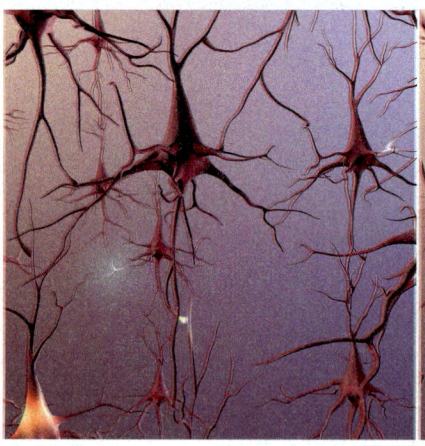

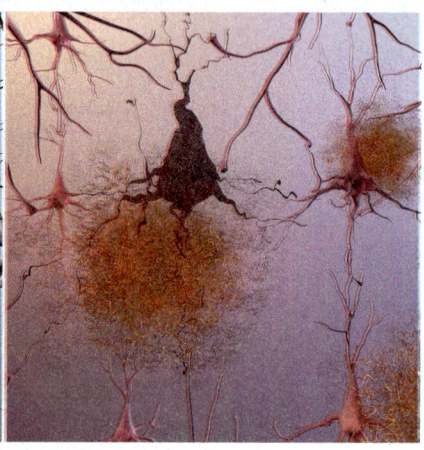

Having had a stroke substantially raises the odds of Alzheimer's, as does having diabetes. But the biggest known risk factor is one people can do nothing about: a family history of the disease. People carrying one copy of variations in a gene called APOE are at about a threefold risk, while those carrying two copies have a whopping 12-fold risk. But it appears possible now that many genes, perhaps even a hundred, may each confer a tiny increased risk of developing Alzheimer's. Still, Rudy Tanzi, an Alzheimer's geneticist at Harvard Medical School, is hopeful that by 2020, a screen of a person's genome will reliably estimate his or her risk of developing the disease. Ideally, the people at highest risk will then have scans or spinal fluid or blood tests regularly to detect accumulating beta amyloid. Once it is seen, they'll begin taking the equivalent of a "statin for the brain" to reduce the load. And Alzheimer's will become as preventable as heart disease is today.

AUTISM: Fighting it in toddlerhood, with play

It was a breakthrough that Lynn Locke of Lodi, Calif., calls "the greatest joy in our lives." After her son Colson, then 3 years old, participated in a year of novel therapy for autism, the brain disorder that inhibits children's ability to communicate and develop relationships, he finally spoke his first word: "Up." A year later, Colson "talks a mile a minute" and is enrolled in a regular preschool class.

Scientists are racing to find autism's causes and

The nerve cells in the healthy brain, at left, form a dense "neuron forest." The brain at right shows Alzheimer's disease.

the drugs that may help treat the disorder, which now afflicts 1 in every 110 children and 1 in 70 boys, according to the Centers for Disease Control and Prevention. But it's behavior therapies like the one that helped Colson that are giving families immediate hope. His success came as part of research at the University of California, Davis Medical Center testing the Early Start Denver Model. The model holds that children as young as 12 or 18 months can improve significantly with more than two dozen hours a week of intensive therapy emphasizing interactive play; often, interventions don't begin until preschool or kindergarten and are more focused on practicing speech and modifying behavior. The goal is to "reduce autism symptoms or prevent them from ever developing," says Geraldine Dawson, a professor of psychiatry at the University of North Carolina-Chapel Hill School of Medicine who helped develop the model, and lead author of a 2009 study of it. Dawson is also chief science officer for Autism Speaks, a nonprofit that funds research. Her report noted that the IQs of children who received Early Start therapy rose on average by close to 18 points, compared to 7 points in the group receiving traditional autism therapies. The Early Start group's listening and understanding skills rose by about 19 points, almost twice the level of improvement in the control group.

The Early Start therapy is built around activities that require the autistic child to make eye contact, take turns, and communicate. Colson and his therapists, parents, and siblings played board games, sang and danced, and marched around the house. When Colson wanted something, he learned to ask for it by pointing and making the sound of a T (his therapist's first initial). He made an M sound to get his mom's attention. Within a few months, he was addressing people by name. Sally Rogers, the UC-Davis psychologist behind the model, is now launching a study of play therapies aimed at infants.

Meanwhile, the exact causes of autism remain a mystery. In the past three years, researchers at Yale, UCLA, and Johns Hopkins have all concluded that it's a disorder of the brain's synapses; the molecules active in an autistic person's synapse don't function properly. A 2009 study noted that the number of autism diagnoses rose 57 percent between 2002 and 2006, which signals to many experts that new environmental factors may increase the odds of developing these dysfunctional synapses. Research from the California Department of Public Health has indicated that advancing maternal age may play a role. Prematurity has been implicated, as has the timing of pregnancy: A recent study found that a second child conceived within a year of an older sibling's birth was more than three times as likely to be diagnosed with autism as children conceived more than three years after the birth. A child's exposure to pesticides and parents' medical conditions, including type 1 diabetes and rheumatoid arthritis, also are under study as possible factors.

While the 1998 study linking the MMR vaccine to autism has been completely discredited, Dawson does not want to dismiss concerns about vaccines entirely. Autism Speaks earmarks 2 percent of its research budget to vaccine studies. The National Institutes of Health, too, has called for more research, given that certain children appear to be more vulnerable than others to vaccine side effects.

BRAIN CANCER: New and better weapons

Each year, more than 22,000 people nationwide develop brain cancer, and more than half die, usually within 15 months. But researchers say they are now poised to make major breakthroughs. "Our understanding of brain tumors, and the way we think about treating them, has changed dramatically after decades of very little progress," says Susan Fitzpatrick, vice president of the James S. McDonnell Foundation in St. Louis, which funds brain cancer research. "We're starting to make headway against a very difficult type of tumor."

Typically, brain cancer is treated with surgery, radiation, and chemotherapy. All present challenges, says Howard Fine, chief of the National Cancer Institute's neuro-oncology unit, because of the brain's

> The latest CDC figures reveal that autism afflicts **1 IN EVERY 110 CHILDREN** and 1 in 70 boys

sensitivity. "Most other organ systems have some potentially expendable normal tissue. You can't just remove half the brain," he says. Radiation inflicts toxic side effects, and while certain drugs effectively kill cancer cells, they kill normal cells as well. The brain is protected by a natural defense system called the blood-brain barrier, which keeps toxins out but sometimes prevents drugs from entering, too.

Now, spurred by genomic research, cancer specialists are beginning to understand the physical and chemical properties that predict which drugs will break through the blood-brain barrier. Therapies that home in on specific tumor cells and spare healthy ones are being developed also; one such drug, Avastin, approved by the FDA in 2009, was the first new glioblastoma treatment in more than a decade. Avastin works by curbing the growth of new blood vessels that supply blood to tumors. "This is a tremendous advance," Fine says. "The response rate with most of our drugs is less than 5 percent. With Avastin, we're seeing upwards of 70 percent."

Imaging is improving, too, leading to better

monitoring and new surgical techniques. Using functional magnetic resonance imaging to map brain activity in the area surrounding a tumor, for example, surgeons can minimize harm. "We're able to do surgery and remove tumors in areas of the brain that we previously couldn't even think about touching," Fine says. A few institutions, including the University of California, San Francisco Medical Center (story, Page 82), are testing a technique that causes even single cancer cells to make their presence known by glowing under fluorescent light.

Research efforts are also targeting stem cells, which have been found in a number of other cancers and are suspected of manufacturing new tumor cells like little factories. Several years ago, researchers discovered that brain tumors contain stemlike cells that can proliferate and self-renew. Coming up with treatments that will defeat these cells might be one key to dramatically improving long-term survival.

CONCUSSION: A heightened respect

Not so long ago, once the confusion, amnesia, dizziness, and nausea from a concussion subsided, the child went back to school, the athlete returned to play, the soldier was sent again into battle. But concussions are no longer being treated quite so lightly, given a growing body of evidence that "mild" traumatic brain injury can result in serious disabilities and fatalities. When the gelatinous brain is subjected to acceleration-deceleration and rotational forces, the metabolism of brain cells is upset and they can no longer function. And it turns out that "second impact syndrome," or a second concussion before the first has had a chance to heal fully, creates a significantly greater risk of dying. Those who survive face long-term consequences.

Indeed, they may eventually suffer from the one form of dementia that's entirely preventable. Ann McKee, an associate professor of neurology and pathology at Boston University who has performed over 50 autopsies on the brains of athletes from age 18 to 83 who played contact sports, has seen evidence of two types of microscopic damage resulting from multiple blows to the head over time. Chronic traumatic encephalopathy, an Alzheimer's-like degenerative disease, strikes the front of the brain, which houses intellect, judgment, learning, and emotions. Affected athletes are in their 30s or 40s before symptoms appear: erratic behavior, staggering gait, amnesia, depression. McKee found a significant indicator of CTE—the buildup of a cell-killing protein called tau—in the brains and spinal cords of deceased athletes and now hopes to identify CTE in the living. Last August, McKee and her colleagues discovered a motor neuron disease similar to amyotrophic lateral sclerosis in three out of 12 deceased athletes, the first evidence that repeated blows to the head might also affect voluntary muscle movements.

How much do parents of athletes need to worry? Researchers at Purdue University outfitted the helmets of 21 high school football players with sensors and found that kids were taking as many as 1,600 hits to the head in a season, registering in many cases some 100 Gs of force (a roller coaster experience equals about 5 Gs). To their surprise, four students who had never actually had a concussion but had taken comparatively more and milder blows to the front and top of their heads showed more cognitive impairment than concussed teammates who took heavy hits mostly to the side of the helmet.

Military wisdom on the topic may soon make it easier to diagnose mild traumatic brain injuries before symptoms appear. About 320,000 troops in Iraq and Afghanistan have suffered concussions from the shock waves generated by roadside bombs and rocket-propelled grenades. So the U.S. Army teamed up with Florida-based Banyan Biomarkers and developed a blood test that accurately diagnosed brain trauma in 34 people with a mild concussion. The test looks for two proteins that enter the bloodstream when the brain is injured. Researchers plan to try it next on 1,200 patients. In addition, the Department of Defense began using

Acceleration-deceleration and rotational forces UPSET THE METABOLISM of brain cells

eye-tracking goggles last year to measure how well an injured soldier is able to pay attention, the primary cognitive function impaired after a concussion. Someone wearing the goggles watches a circling dot for 30 seconds; ability to focus while tracking the target is measured, and a score is generated to quantify the severity of an injury.

There is optimism, too, about the severe brain traumas that cause a form of coma called persistent vegetative state, in which a person is awake but not aware. Only 3 to 7 percent of people in this state typically recover. But new tactics to jump-start these patients' brains are showing promise as part of the "emerging consciousness" program at the Department of Veterans Affairs. The focus is on combining nursing and rehabilitation services, an individualized therapy program, intensive case management, and psychological support services and education for families and caregivers—a regimen that goes beyond standard care. The program also relies on new approaches such as the use of stimulant drugs like Ritalin, bromocriptine, and modafinil, which appear to galvanize the brain. The VA reports that its multi-pronged program has brought nearly 70 percent of patients back to consciousness.

PARKINSON'S DISEASE: Beyond drugs

Drug treatment for Parkinson's, an incurable neurodegenerative disease, typically targets the characteristic motor disturbances such as tremors, slowed movement, and rigidity by compensating for insufficient dopamine in the brain. As the disease progresses, however, the drugs are less effective and side effects are more evident. The therapy also may not address other significant issues associated with Parkinson's, including depression, anxiety, apathy, thinking and memory problems, and irregular sleep. With the disease now affecting perhaps a million Americans—a number likely to grow as the population ages—complementary strategies are attracting growing interest. Animal studies suggest, for example, that vigorous exercise may have neuroprotective effects, slowing the brain's loss of dopamine.

Gait and balance can also be big problems for Parkinson's patients. At Washington University School of Medicine in St. Louis, researchers tested the impact of tango lessons on two groups with moderate symptoms of the disease. In one group, participants danced alone. In the other, they danced with partners. While both groups showed significant improvement, those who tangoed with partners indicated more interest in continuing—the only way that the benefits could be sustained. The study, published in the May 2010 issue of the journal *Neurorehabilitation & Neural Repair*, points to the need to get patients involved in physical activities that they can maintain over the long term.

Researchers are also exploring why certain people experience temporary relief from Parkinson's symptoms when they respond to specific cues. For example, some patients can overcome their hobbled movements after hearing familiar rhythms. Others, barely able to walk, can dance when prompted, or can sing effortlessly, even if they have difficulty speaking. For reasons not well understood, auditory cues seem to reawaken, or somehow reconnect, brain circuits that otherwise aren't working as they should.

A team of Italian researchers took this idea a step further, putting people with moderate Parkinson's into a theater group so that they were compelled to take control of their movements and emotions. The

> Animal studies suggest that **VIGOROUS EXERCISE** may slow the loss of dopamine in Parkinson's patients

three-year pilot study, reported in 2010, found that participants became more confident and motivated. And on a range of measures from mood to mobility, they outperformed a control group given regular physical therapy.

Why such approaches seem to succeed, even if only temporarily, remains largely a mystery. Rigorous studies on them have not been funded to the same extent as efforts to find better drugs, research the feasibility of putting corrected copies of defective genes into the brain, or implant brain-stimulating electrodes, for example. Some experts estimate that the electrodes could help 10 percent of Parkinson's patients (story, Page 82). But evidence is mounting that many patients benefit from music and movement therapy, at least in the short term.

SCHIZOPHRENIA: Striving for early detection

This debilitating mental illness has been diagnosed for a century based on symptoms that patients report—the hallucinations, the inability to feel or show emotion, the social withdrawal. That's a major reason there's typically nearly a nine-year gap between the first signs of trouble and the start of treatment. Now, researchers say they are on the verge of developing diagnostic techniques that rely on biomarkers like blood proteins or genes, which could mean much earlier diagnosis and a better outcome. "Our goal is to find something objective," says Stephen Glatt, a psychiatric geneticist at the State University of New York Upstate Medical University in Syracuse. "It holds a lot of potential for people suffering from this condition." Early detection might also identify young people predisposed to schizophrenia in time to head off the worst of its effects.

The news is less good on the treatment front. Standard antipsychotic or neuroleptic drugs, which change the balance of chemicals in the brain and help control symptoms, can lead to sleepiness, weight gain, movement problems, and muscle contractions. While newer drugs called atypical antipsychotics come with fewer side effects, they aren't significantly more effective, Glatt says.

Meanwhile, other ways of managing schizophrenia are growing in popularity. Psycho-education teaches patients and their families about the illness and how to best avoid a relapse. Another approach, problem-oriented personalized psychotherapy, provides guidance on tackling everyday problems. Patients can learn to change their problematic thoughts with cognitive behavior therapy, and cognitive remediation uses computer-assisted training exercises to work on memory, attention, and problem-solving. Combined with medication, these options can make a big difference, experts say.

But with mental health and hospital resources scarce, it doesn't always happen, notes Jeffrey Lieberman, director of the New York State Psychiatric Institute in New York City. "There's a lot more that could be done in this field simply by applying what we already know."

STROKE: Widening the window for treatment

Time is brain, as the saying goes, and the push continues to make the most of those critical first hours after a stroke. The standard treatment for dissolving clots in ischemic strokes is intravenous tPA, or tissue plasminogen activator; administered within three hours, it can greatly increase the odds of recovery. But only one third of candidates make it to the hospital that quickly, and since the treatment isn't appropriate for a hemorrhagic, or bleeding, stroke (it would raise the risk of bleeding), it can't be started in the ambulance.

One promising technique now under study in California is to have emergency medical technicians administer magnesium sulfate (which dilates blood vessels and blocks the calcium buildup responsible for cell death) while en route to the hospital. "The hope is that more brain will be saved with this therapy," says Walter Koroshetz, deputy director of the National Institute of Neurological Disorders and Stroke. Meanwhile, evidence from Europe indicates that tPA can be administered effectively in some people up to four and a half hours after a stroke, and injecting tPA by catheter directly to the clot through an artery in the groin opens the window up to six hours. In some cases, it's possible to remove the clot using tools threaded through blood vessels in the brain; the Merci Retriever System, a coil-shaped device that pulls out clots, and the vacuum-like Penumbra System, which sucks them out, can be used up to eight hours after the first symptoms.

Once the damage is done, how to bring function back? Though many physicians still doubt motor skills can be regained much beyond six months out, the latest research indicates otherwise. One avenue of study suggests that high-intensity repetitive exercise, aided by robots or people, can significantly improve arm functioning even several years later. Transcranial magnetic stimulation also appears to help partially paralyzed patients as much as three years later. And researchers at the Toronto Rehabilitation Institute have reported that just eight sessions of Wii Tennis and Wii Cooking Mama—in which a player simulates peeling, cutting, and slicing—gave motor function a measurable boost.

By using stem cells to stimulate neuron and blood-vessel growth, researchers hope to someday repair stroke-injured areas. Last November, Scottish researchers injected fetal stem cells into the brain of a man in his 60s who had suffered a debilitating stroke 18 months earlier. The man will be monitored over two years to see if he gets better; 12 additional patients will get the therapy this year. •

By Brian Vastag, Kathryn Roethel, Angela Haupt, Donna Banks, and Keith Sinzinger

> Early detection might identify young people at risk of **SCHIZOPHRENIA** in time to head off the worst effects

HEALING THE BRAIN

What if the Gloom Won't Lift?

WHEN DRUGS DON'T OFFER ANY RELIEF, ELECTROCONVULSIVE THERAPY IS ONE OPTION

The era of easy antidepressants has not changed this reality: Successfully managing depression is hard. What next, if a drug doesn't work for you?

A couple of years ago, a University of Kansas student, confined to bed by intense sadness, exhaustion, and headaches, found herself considering suicide. Desperate after a years-long struggle with depression, she sought a treatment she had once viewed as extreme: electroconvulsive therapy, or ECT. After a few sessions, "I literally went from almost unable to function—feeling suicidal—to a 180-degree change," she says.

That student counts as one of the many people chronically in depression's grip who, disappointed by drugs, have been finding some relief in alternative therapies ranging from exercise to various forms of high-tech brain stimulation. While antidepressants are the most widely prescribed drugs in the United States today, thanks largely to the arrival of Prozac and other effective options with fewer side effects, a groundbreaking 2006 trial known as STAR*D found that only about one third of people found total relief with the first drug prescribed, and around a third were not helped even after trying several drugs and combinations. Research published last year revealed that many people with the most severe symptoms *can* be helped by drugs, though those with mild or moderate depression might as well take a sugar pill. ECT, which has been controversial since the days when it was performed without anesthesia and sometimes without proper consent, has evolved considerably in recent years; by inducing a seizure, it is thought to reset dysfunctional brain circuitry. It "is the most effective and rapidly acting treatment for severe depression," says Sarah Lisanby, chair of the psychiatry and behavioral sciences department at Duke University School of Medicine and a leading brain stimulation researcher.

Because ECT is an invasive therapy that involves anesthesia and often memory loss, people suffering from unrelenting depression are steered to other approaches first. These might include continued medication—though getting a response can take considerable work. Steven Hollon, professor of psychology and a depression researcher at Vanderbilt University, is concerned that family practitioners, who have become much more comfortable writing prescriptions for the newer antidepressants, don't offer enough follow-up. It can take six weeks for an antidepressant to kick in; many people simply give up, especially if the new drug comes with, say, headaches or an upset stomach. "That can be asking a lot of a person," says Matthew Rudorfer, psychiatrist and assistant chief of the adult treatment and preventive intervention research branch at the National Institute of Mental Health. It may well be, he says, that side effects would subside, or that switching drugs or adding a second one can work.

Or perhaps a dose of therapy is called for. An August 2009 report in *Archives of General Psychiatry* revealed that people on antidepressants are less likely to also be in therapy than they once were—about 20 percent in 2005, down from nearly 32 percent in 1996. But some evidence suggests that chronic depression may respond more readily to medication plus therapy than to either alone. And one arm of the STAR*D trial showed that turning to cognitive behavioral therapy, or CBT, after a first drug fails works about as well as trying a second medication. The focus is tightly on correcting the negative or catastrophic thought patterns ("I'm such a failure," "I'm not worthy of being loved") that so often stoke depression. Some intriguing work in neuroimaging has shown that CBT "not only works to relieve symptoms but is also associated with brain function changes," says Madhukar Trivedi, a lead researcher on STAR*D and a professor of psychiatry at University of Texas Southwestern Medical Center.

Clearly, the way people think matters. "In a depressed person's mind, thoughts tend to be overly pessimistic, overly harsh in regard to how the world works," explains Robert DeRubeis, a psychologist and a depression and CBT researcher at the University of Pennsylvania. "Our behaviors follow from the judgments we make" and often just deepen feelings of woe. A depressed person may decide not to attend a party, for example, because he believes no one will talk to him. But with a therapist's probing, he might examine how realistic that belief is and realize he has the power to start the conversational ball rolling. Some research suggests CBT may have a more lasting effect than antidepressants after treatment ends, perhaps because people have mastered the strategies that keep them from getting depressed, says Hollon.

Experience matters. It's important to find a well-trained cognitive behavioral therapist. A 2005 study by DeRubeis and Hollon comparing 16 weeks of drugs, CBT, and a placebo found a response rate of 58 percent in both the drug and CBT groups—but also that the level of therapist expertise might affect CBT's success rate. How best to find a practitioner? Start by inquiring at a nearby academic medical center or by searching the Academy of Cognitive Therapy's website (*www.academyofct.org*). And give it two to five sessions before doing a gut check, says Hollon.

HEALING THE BRAIN

Someone referred for electroconvulsive therapy would typically start out with several sessions per week, then would taper off and stop over a period of months. Short-term memory loss is the main concern with ECT, and it's not uncommon. The effect usually wears off after treatment ends, but some information may never return—that retirement party or graduation ceremony you attended between sessions, for example. Some patients claim to have experienced more substantial problems, which may be a consequence, say, of using more current than was necessary. ECT has changed significantly as understanding has grown about how to minimize memory side effects, says Rudorfer of NIMH. He says it "has much lower risk than decades ago—though the risk is not zero." Technique matters, including an ability to reach just the amount of electrical current needed to induce seizures, which can differ among patients; the placement of the electrodes on the head; and the type of stimulation used (brief or ultrabrief pulse causes the fewest cognitive deficits while an older type, sine wave stimulation, produces significantly more). Cognitive problems are considerably less pronounced when the electrodes are put on one side of the head instead of both, but the one-sided approach is not as effective in some people. Critics of ECT have insisted that it causes brain injury, but studies in humans and animals have not corroborated the claim, says Rudorfer.

Another caveat: The benefits don't necessarily last. One study showed that 84 percent of patients had relapsed six months after the treatments ended without any "maintenance therapy" (such as drugs, which may help after ECT even if they failed before, or less frequent ECT sessions). Still, other research has shown that after a successful course of ECT with some form of maintenance therapy, about 46 percent of patients remained well six months out. The college student eventually began trying to manage her recurring depression with a combination of therapy, medication, and lifestyle changes—more exercise and sleep, and light exposure, for example—that she learned about from *The Depression Cure: The 6-Step Program to Beat Depression Without Drugs*, a 2009 book based on research by her psychology professor at Kansas, Stephen Ilardi.

Get moving. Indeed, a growing body of research suggests that regular exercise, at least, might be a smart prescription to try or to add to drugs or therapy. It appears to promote a good, stable mood by reinforcing self-confidence and a sense of control over one's health, says Andrea Dunn, a Colorado behavioral science researcher and a principal investigator for a pilot study exploring the impact of regular exercise on depressed adolescents. A possible mechanism: Exercise creates new neurons, she says, bolstering connectivity in the depressed brain, which often operates with a deficit of connections.

The oft-quoted success rate for electroconvulsive therapy—that it brings remission to 80 percent of people who try it—was arrived at under the rigorous setting of a clinical trial. A 2004 study of ECT success rates in the community hospital setting put the number between 30 percent and 47 percent. The discrepancy probably reflects variation in doctors' techniques, additional complicating illnesses, and the fact that some patients stop ECT early because of side effects, says Lisanby.

In the face of ECT's shortcomings, newer brain stimulation treatments are being explored. But

> It can take weeks for an antidepressant to kick in; **MANY PEOPLE** give up, especially if they suffer side effects

none is widely available, and how well each works and for whom is not yet known. Transcranial magnetic stimulation, which has been cleared by the Food and Drug Administration for depression, targets neurons in areas of the brain involved in mood by placing a magnetic coil on the head. Unlike ECT, it does not result in a seizure or require anesthesia, nor does it cause memory loss. But it is hardly available in every community and may not work well for patients with more severe cases. In vagus nerve stimulation, a device surgically implanted in the chest stimulates the brain by shooting electrical pulses into the vagus nerve in the neck. "To many in the field, the jury is still out" on its effectiveness, says Rudorfer. Deep brain stimulation, approved for movement disorders including Parkinson's disease, is available for depression in research trials. It involves surgery to place electrodes into the brain. The good news, says Rudorfer, is that most people will respond to a mainstream treatment—as long as they persevere. ●

By Sarah Baldauf

DID YOU KNOW...

Taxi drivers in London spend years memorizing and reciting routes before they are permitted to get behind the wheel. Brain scans have shown that the hippocampus, the navigation and memory center of the brain, is on the large side in these cabbies, possibly a result of their street smarts.

SOURCE: STUDIES IN 2000 AND 2006 BY ELEANOR MAGUIRE, UNIVERSITY COLLEGE LONDON

HEALING THE BRAIN

Three Patients, Three Operations

TAKE A FRONT-ROW SEAT IN NEUROSURGERY AT THE UCSF MEDICAL CENTER

SAN FRANCISCO—If a person suffering from the tremors and rigidity of Parkinson's disease chooses brain surgery, she will most likely need to go off her medications and be awake during the procedure so doctors can gauge her ability to speak and move as they maneuver. That's been the uncomfortable reality for people who have experienced deep brain stimulation, a state-of-the-art technique that implants electrodes to calm erratic nerve signals and ease symptoms. But at the University of California, San Francisco Medical Center, she has an even more cutting-edge option: She can stay on the meds and sleep through the operation inside an MRI machine while surgeons drill into her skull and use imaging to guide the electrodes safely into place.

This groundbreaking approach to deep brain stimulation (DBS), pioneered here and headed soon to the Cleveland Clinic, is one of the many examples of envelope-pushing going on at UCSF. The medical center, No. 5 in neurology and neurosurgery in the latest *U.S. News* Best Hospitals ranking, performed 3,479 neurosurgeries last year. While most are fairly standard fare, such as removing benign tumors and easing seizures, people from around the world come here to put their brains and their hopes in the hands of surgeons testing medical frontiers. One is using a dye

Photography by Daryl Peveto-Luceo for *USN&WR*

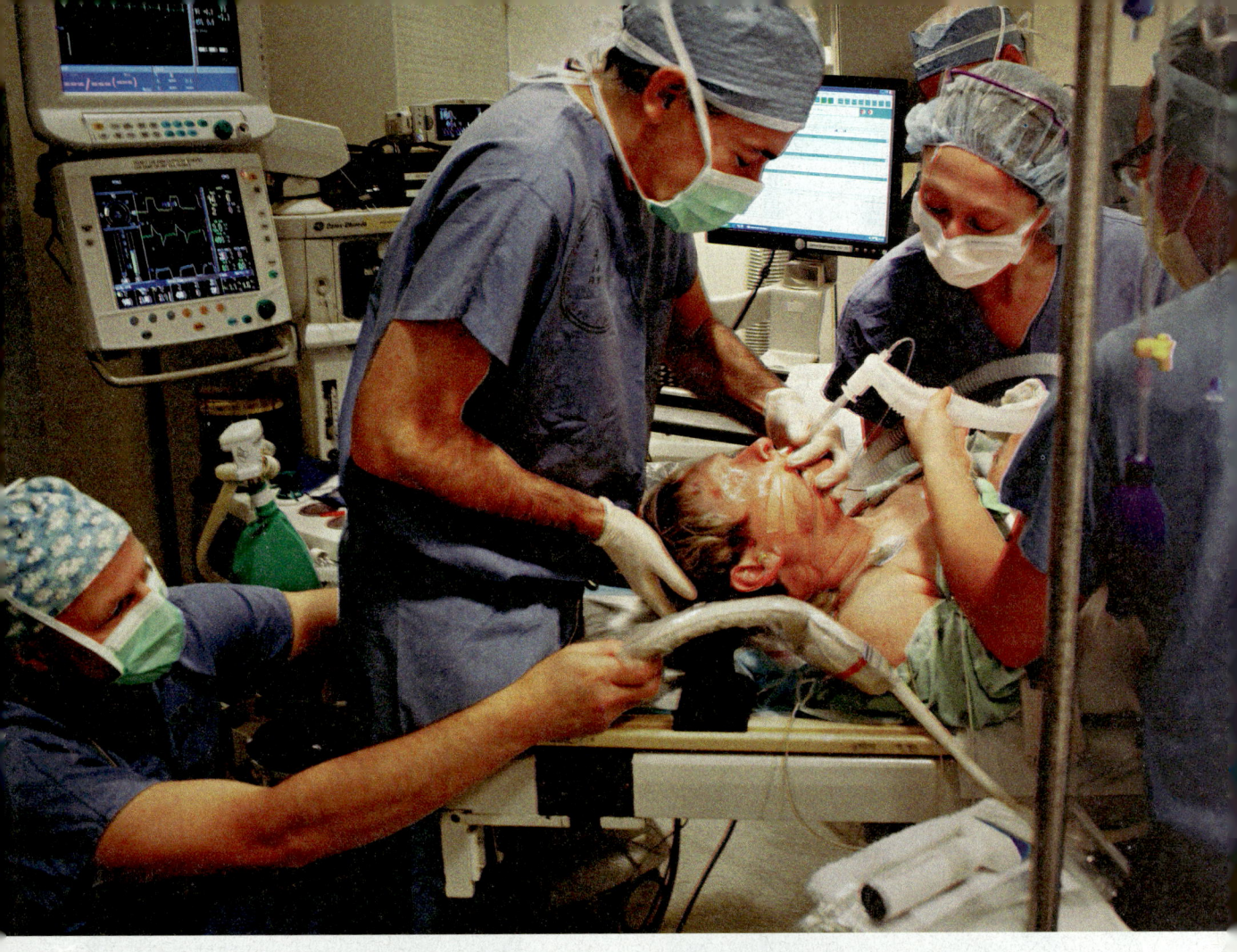

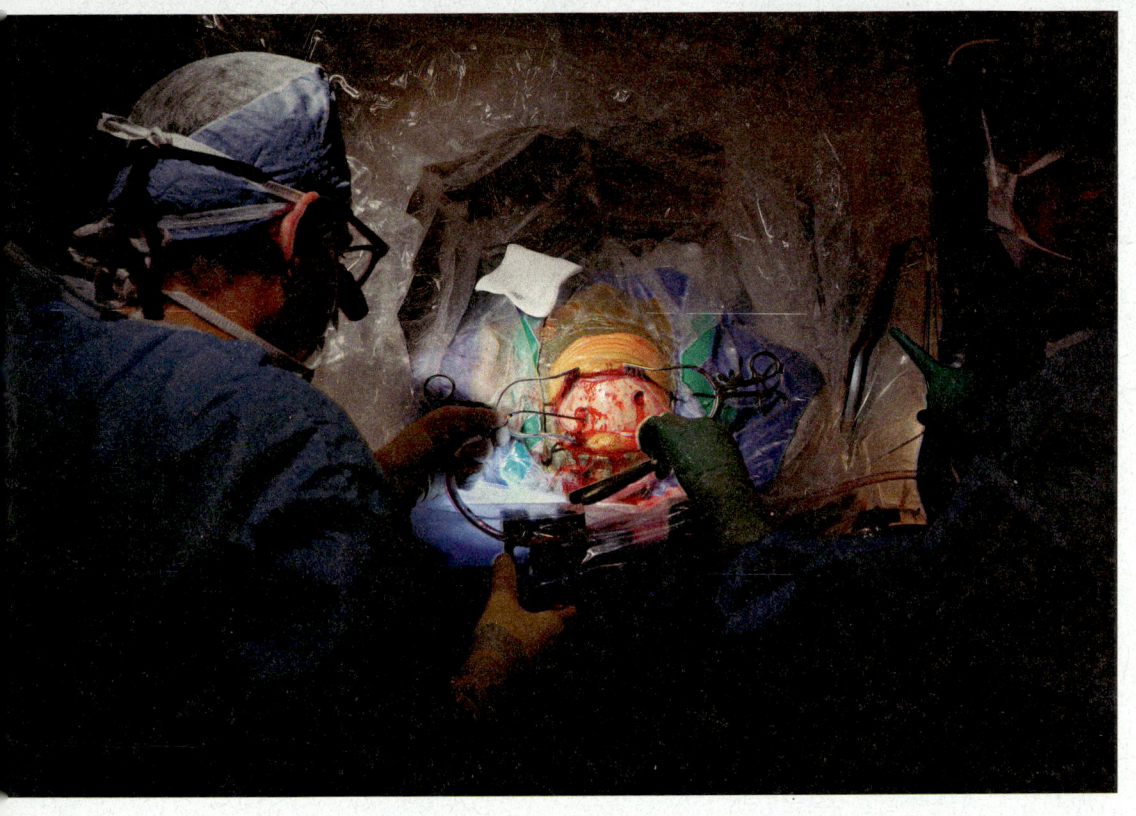

Parkinson's patient Linda Sharp (top left) chose deep brain stimulation using interventional MRI, allowing her to sleep through it. As the MRI room was being prepped as an OR (opposite bottom), surgeon Philip Starr readied her for surgery (above), which involved drilling into her skull and using imaging to thread electrodes into place (at left).

SPECIAL EDITION 83

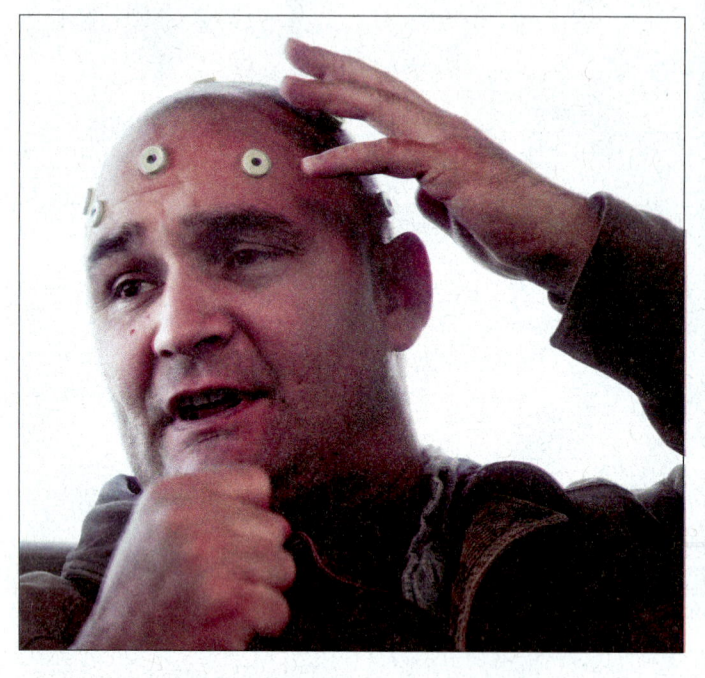

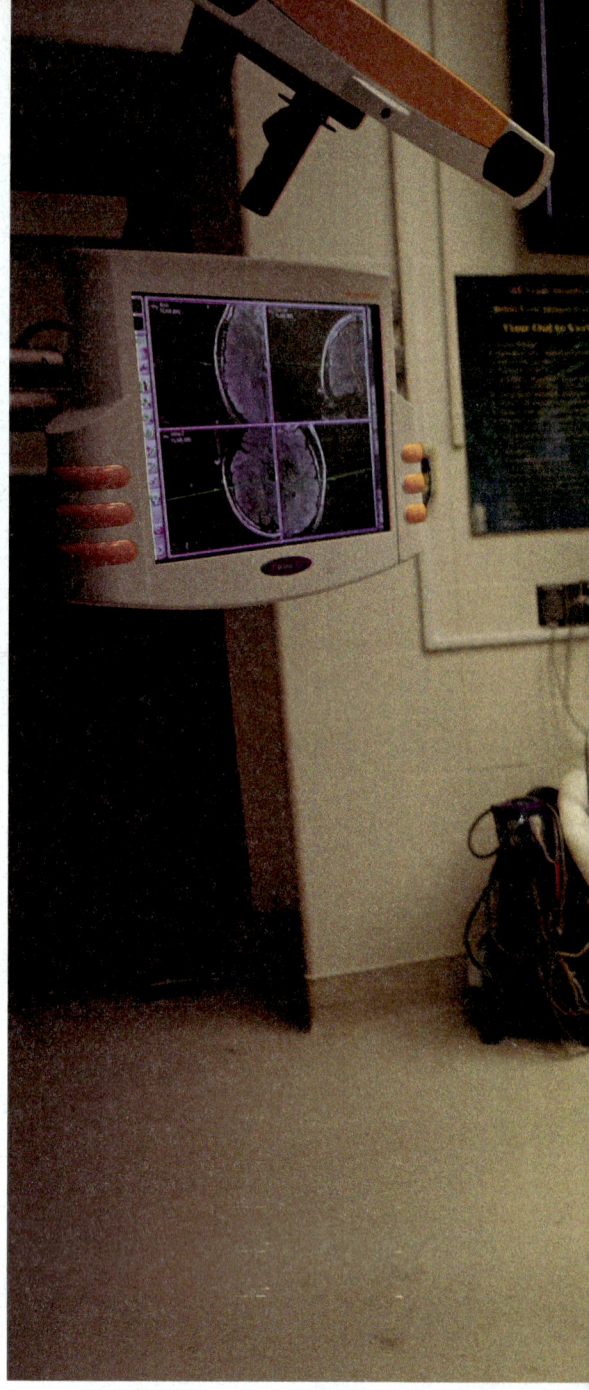

After Daniel Sheafer (pre-surgery, top) had his tumor removed, surgeon Edward Chang gave fiancee Gina Meldrum (bottom) the good news.

that makes even single malignant cells glow orange under fluorescent light so it's possible to better target them for extraction. Another research project, also in the Parkinson's realm, is investigating the potential of infusing corrected copies of defective genes into the brain; clinical trials could begin as early as this year. One surgeon is mapping electrical activity when a person who is undergoing "awake-brain" surgery speaks; the goal is to create a voice prosthetic for people who can't talk. *U.S. News* recently spent two days at UCSF to watch the neurosurgeons in action.

An MRI room isn't normally set up for surgery, but the imaging center has been transformed today into a bona fide operating room. By readying a set of non-magnetic titanium instruments (so the giant magnets at the core of the imaging system don't tug them out of the surgeon's hands) and draping sterile sheets around the opening of the MRI tunnel, the surgical team has prepared the suite for Linda Sharp's deep brain stimulation on this chilly San Francisco morning in January.

Sharp, 71, began her journey here about 12 years ago, when her left leg began to fail during walks to her gym in Los Osos, Calif. In 2001, she was diagnosed with Parkinson's and subsequently developed dyskinesia, which produces uncontrollable bouncy movements in her legs. Meanwhile, neurosurgeon Philip Starr was brainstorming a better way to implant electrodes for DBS. About 50,000 Americans are diagnosed with Parkinson's each year, he says, and about 10 percent of them could benefit from the stimulation. But many are not good candidates for awake-brain surgery because they're too nervous or experience involuntary movements. Taking advantage of the rapidly advancing quality of MRI images, Starr, fellow surgeon Paul Larson, and MRI physicist Alastair Martin developed an "interventional MRI" procedure that allows them to see deep inside the brain as they operate and the patient sleeps. Collaborating with SurgiVision, a Memphis-based company that develops MRI software and hardware to guide electrodes into the brain, the team received Food and Drug Administration approval last July for the new iMRI system. (Elsewhere, MRI-guided surgeries are being tested or used to accurately point a laser or high-intensity focused ultrasound at some tumors and uterine fibroids.)

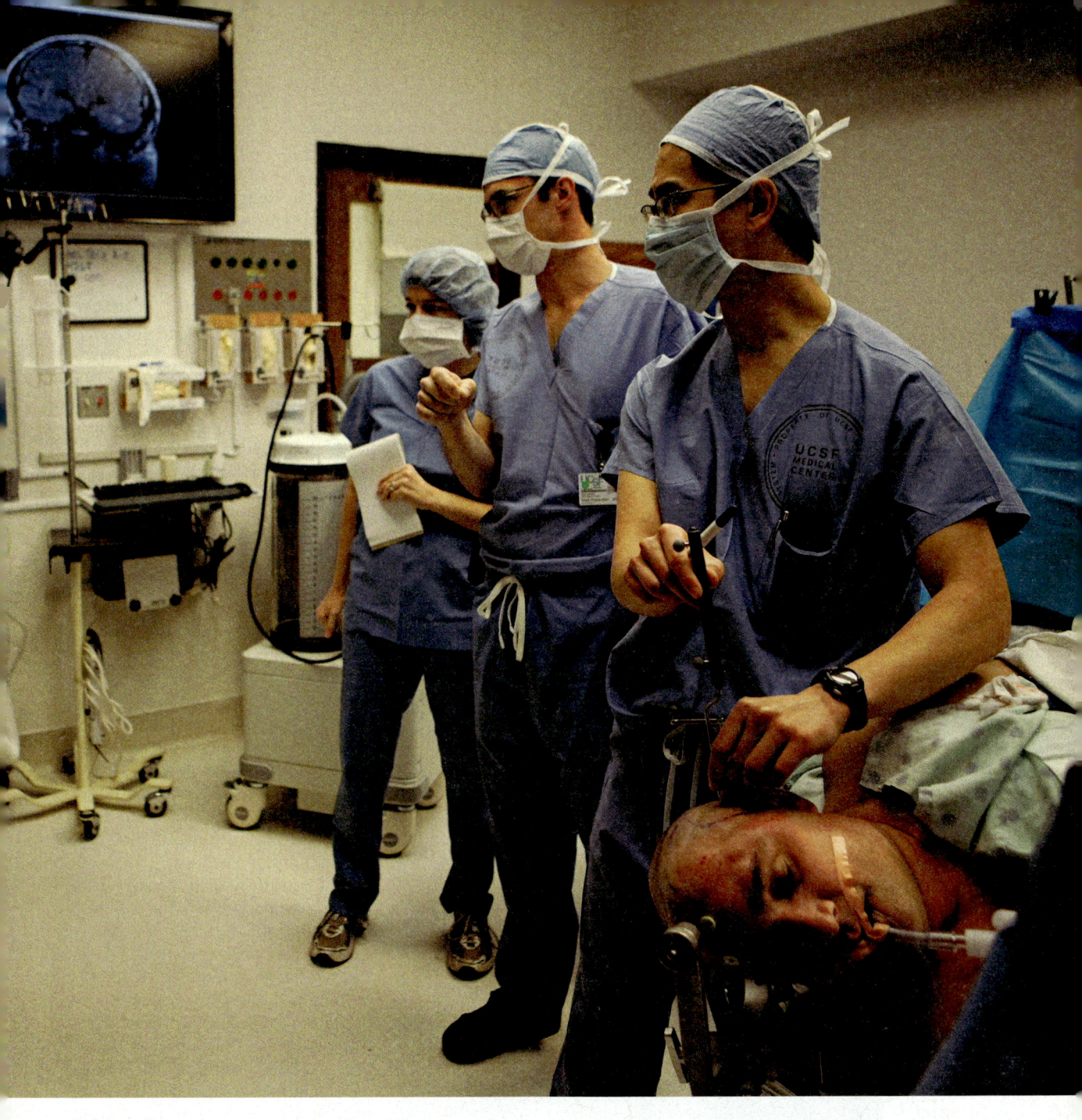

On this morning, Sharp was positioned inside the MRI scanner to capture a pre-surgery image that would determine the safest trajectory for the electrodes, plastic-encased wires as flexible as partially cooked spaghetti. The surgeons then drilled two nickel-sized burr holes in her skull and affixed a plastic frame that would allow them to precisely thread the electrodes, by turning knobs to tweak the angle of entry, into each hemisphere's subthalamic nucleus, the problem zone. Images were taken

Chang (above right) scanned the terrain before opening Sheafer's skull and waking him up to map (and thus avoid) areas key to speech and movement.

throughout the insertion to watch for hemorrhaging and to ensure that the electrodes hit the target.

The night before her operation, Sharp had admitted to feeling apprehensive; one of her friends had suffered speech problems following traditional DBS surgery. But after the roughly three-hour procedure, she came out of recovery smiling and with her speech intact. A week later, a second surgery threaded the ends of the electrodes, which had been coiled under Sharp's scalp for safekeeping, under the skin of her neck and connected them to a device placed under her collarbone that generates electrical pulses. The full benefits should be seen over the next six months, as the pulse is programmed to suit her and her meds are adjusted.

The day after Sharp's procedure, Daniel Sheafer's brain was exposed to the air in a standard operating room. Sheafer, 38, of Modesto, Calif., learned he had a tumor in his temporal lobe last year after painful headaches kept him from working his landscaping and other jobs. "It felt like a cue ball was bouncing from side to side in my head," he recalled at UCSF the day before his operation. The fact that Sheafer suffers from epilepsy complicated matters, says surgeon Edward Chang.

If the tumor were accessible, Chang planned to remove as much of it as he could (iMRI wouldn't be used; the procedure is approved specifically for DBS). He would also try to locate the source of Sheafer's seizures, in the hope of addressing it surgically. Finally, if Sheafer stayed calm during the awake-brain part of the procedure, Chang would take advantage of the patient's willingness to help out with his research: gathering data about the electrical activity that produces speech. Chang would apply sensors directly to Sheafer's temporal lobe to record the location of electrical bursts produced when he repeated certain words.

Chang has been performing these on-brain electroencephalograms, or EEGs, for two years, using the data to refine software algorithms that match electrical signals to sounds like "ba" and "da." The next step is to correlate the signals with sounds as a person thinks them. Eventually, Chang hopes, a device implanted on a person's brain will "read brain activity, do computations, and wirelessly output the information to a computer that decodes it and fits it into a speech synthesizer." Think it, and a computer will say it. While a voice prosthetic will likely take years to perfect, the first human trials should begin this year.

After Sheafer, a devout Christian, had led the team in a prayer, he was sedated so the surgeons could remove part of his skull. Then, to safely get at the wispy, wave-shaped tumor 3 centimeters beneath the surface, Chang needed his patient awake so he could map the parts of Sheafer's brain that control his movement and speech. The surgeon electrically stimulated Sheafer's cortex while asking him if and where he felt tingling sensations in his face. He also asked Sheafer to identify objects in pictures and to count. All the while, eight electrodes kept watch for seizure activity.

After mapping the brain for the tumor removal, Chang applied his speech sensors, and Sheafer repeated words from a list while electrical signals were recorded. On a computer monitor, a yellow line jittered as Sheafer's brain donated its language secrets to science.

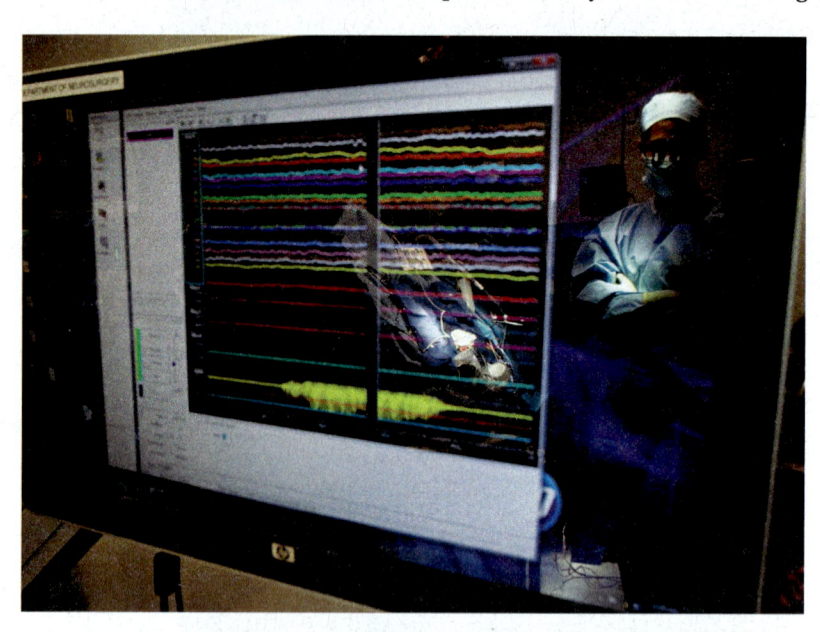

Chang's work on a prosthetic to turn brain waves into speech got input from Sheafer, who spoke as sensors on his brain (right) monitored activity associated with sounds.

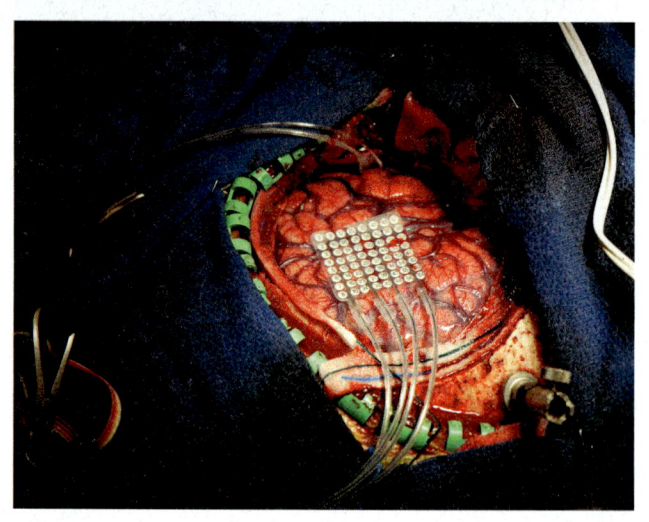

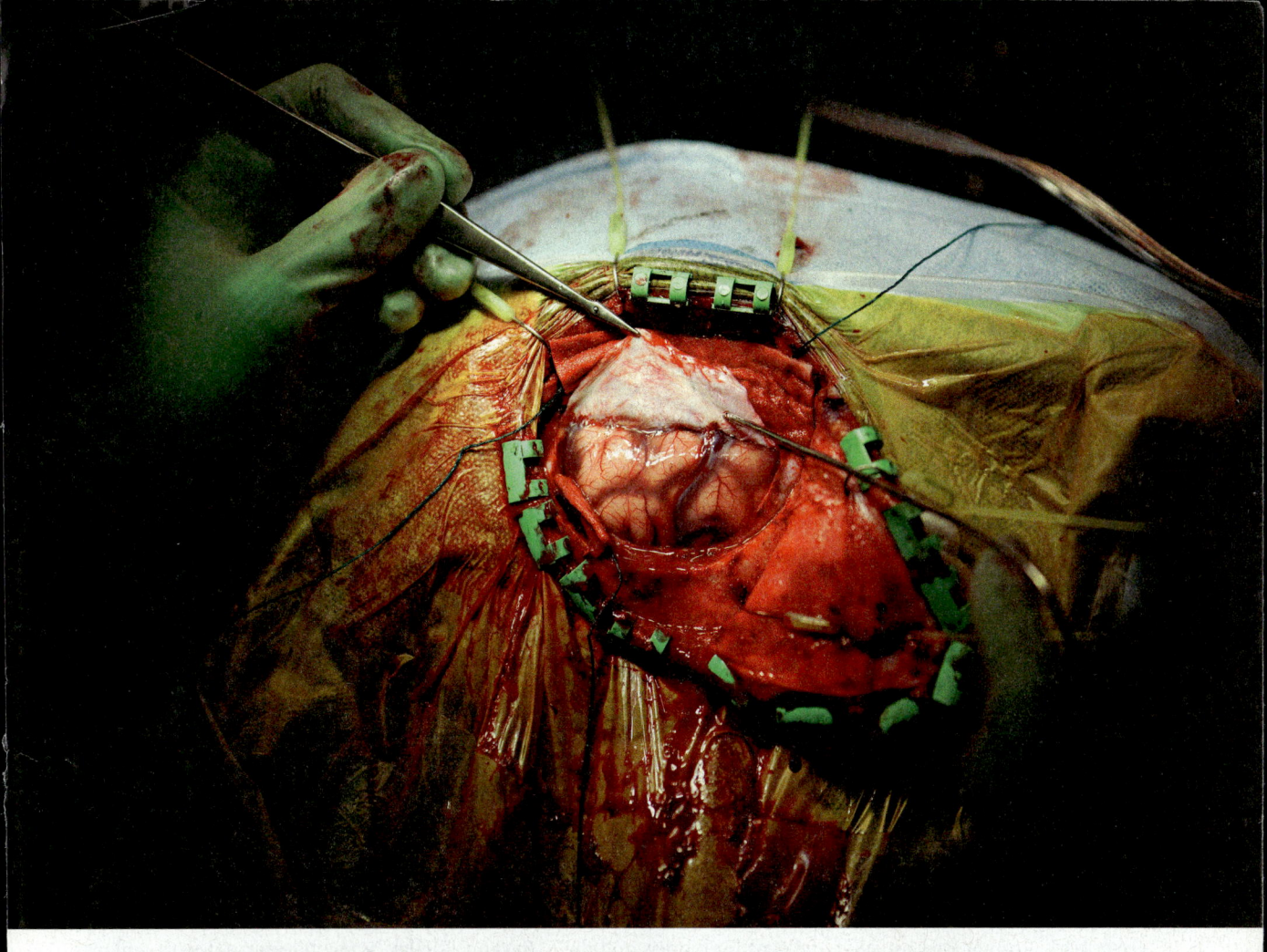

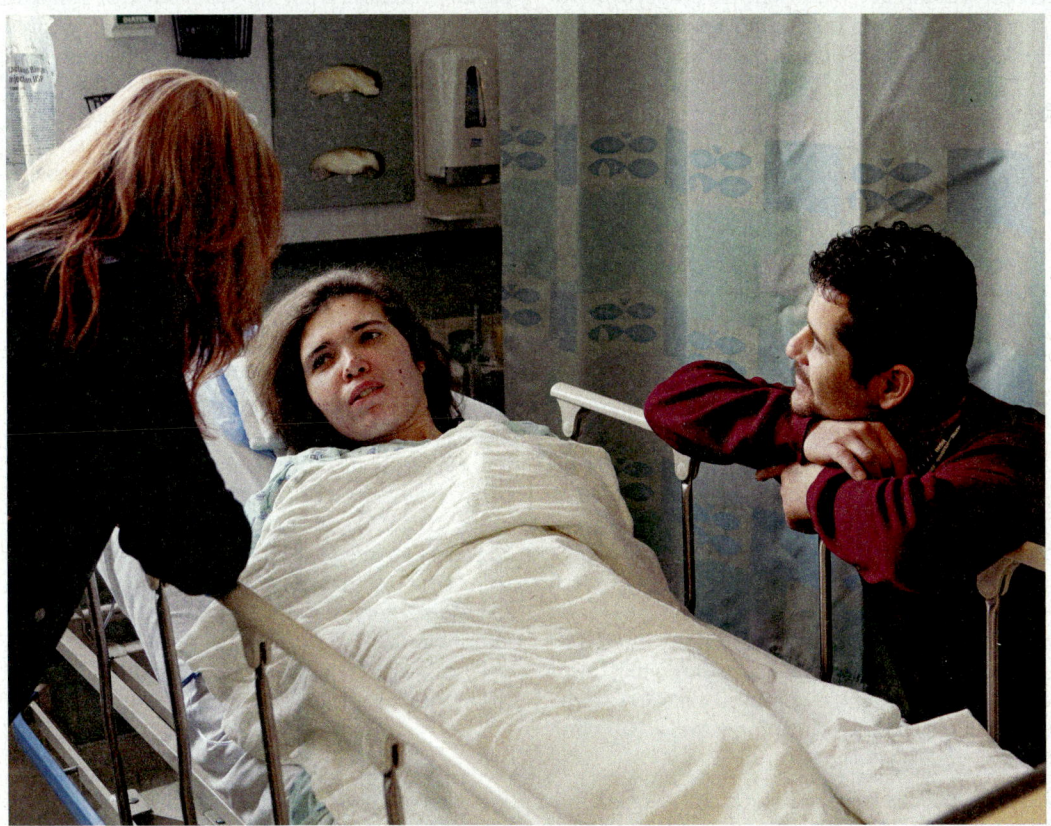

Before surgery (above), Savanna Kelley and her parents, Marisa Martin and Brian Kelley (at left), hoped that her corpus callosotomy would ease seizures so severe that they cause her to collapse. The operation severs many of the fibers connecting the brain hemispheres, disrupting signals between them.

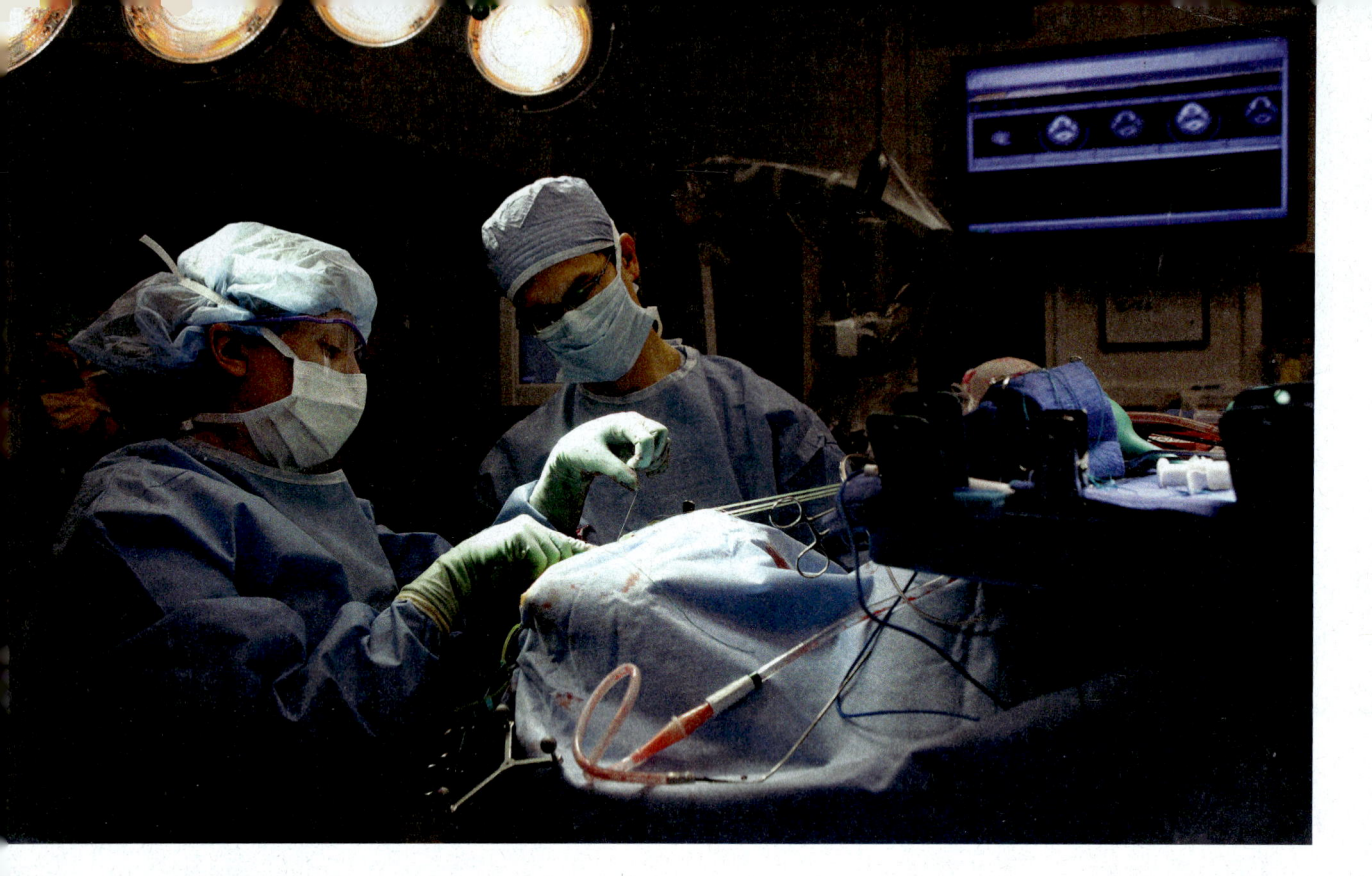

Kelley's surgeon Chang (above right), operating with clinical fellow Ellen Air, estimates that 50 or so corpus callosotomies are performed in the country each year. While not a cure for epilepsy, the procedure alleviates most patients' "drop attacks."

With his brain mapped and data-gathering complete (no luck pinpointing the seizures), Sheafer was again sedated. Chang scooped out an inch-square chunk of lateral cortex, revealing the dark gray tumor below. A portion of the tumor was cut away and sent for biopsy, and the remaining tendrils were removed with a suction tool. Awake again, Sheafer was able to talk and get off the operating table himself before being wheeled to recovery.

Luckily for Sheafer, the tumor turned out to be benign, and it was removed completely. For those with malignant growths, especially ones that sprawl, UCSF Medical Center is working on a technique that could produce much better results. Mitchel Berger, chairman of the department of neurological surgery, is one of a handful of doctors in the United States with FDA approval to do the dye-activated visualization procedure, which has been shown to significantly improve survival. The patient drinks an organic dye called 5-ALA, which targets tumor cells by interacting with their overactive mitochondria. When the cancer cells metabolize the dye and are exposed to fluorescent light, they glow orange, indicating with unprecedented accuracy exactly what a surgeon should remove.

"It's like cutting along the dotted line," says Berger, who has completed more than a half dozen of the procedures and is approved to do 50 within a year. Once a study at multiple centers is complete, says Berger, there's a good chance that the procedure could be broadly approved by the FDA; he estimates this could happen in three to five years.

That afternoon, Chang was back in the OR with Savanna Kelley, a shy 23-year-old with epilepsy who was regularly experiencing grand mal seizures powerful enough to send her to the ground. Kelley, who is developmentally delayed, had recently started a vocational program near her hometown of Lemoore, Calif., but the "drop attacks" had been getting in her way. "She wants to do stuff by herself," says Marisa Martin, Savanna's mom. "It's frustrating for her."

Kelley's procedure, a corpus callosotomy, severs many of the fibers of the corpus callosum, which connects the brain's hemispheres and facilitates communication between them. Afterward, the firestorm that starts on the left side of Kelley's brain won't be able to spread to the other side. She'll still have seizures, says Chang, but they'll be less severe. It usually takes a few weeks to know the effect, he says, but most patients see improvement. UCSF Medical Center does several of the country's 50 or so corpus callosotomies each year, so its surgeons are considered old hands. This is "an old surgery," says Chang. "It's tried and true and it works."

From the patients' perspective, that doesn't make the artistry, or the outcome, any less dramatic. ●

By Kate Greene

MORE @ USNEWS.COM For the country's best hospitals in 16 specialties, visit **www.usnews.com/best-hospitals**.